IITK Directions

Volume 8

IITK Directions is an institutional series of the Indian Institute of Technology Kanpur. IITK Directions aims at presenting original contributions and reviews of important and cutting-edge topics by the faculty members and other stakeholders of IIT Kanpur in a consolidated manner. It is a platform that reflects upon the ongoing research and development activities, major achievements, and evolving trends. Each volume in the series focuses on a particular area of science and technology with chapters written by the faculty members, students and research staff. Each chapter is written in a journalistic tone for the peer group – readability and accessibility being important parameters. The volumes contain a survey of the subject as a whole and the extent of contributions recorded by the Institute faculty and students who participate so extensively in research.

The volumes in this series will be useful to researchers, practitioners, students, industry leaders, government officials and policy makers. Above all, the series serves as platform to publish the most cutting edge research summaries and evaluate the applications and impact of the research work undertaken at IIT Kanpur.

Sudarshan Narayanan · Laltu Chandra ·
Deepika Jhajhria
Editors

Materials, Devices and Systems for Sustainability

Springer

Editors
Sudarshan Narayanan
Department of Sustainable Energy Engineering
and Kotak School of Sustainability
Indian Institute of Technology Kanpur
Kanpur, Uttar Pradesh, India

Laltu Chandra
Department of Sustainable Energy Engineering
and Kotak School of Sustainability
Indian Institute of Technology Kanpur
Kanpur, Uttar Pradesh, India

Deepika Jhajhria
Department of Sustainable Energy Engineering
and Kotak School of Sustainability
Indian Institute of Technology Kanpur
Kanpur, Uttar Pradesh, India

ISSN 2509-6591 ISSN 2509-6605 (electronic)
IITK Directions
ISBN 978-981-96-7294-3 ISBN 978-981-96-7295-0 (eBook)
https://doi.org/10.1007/978-981-96-7295-0

This Springer imprint is published by the registered company Springer Nature Singapore Pte Ltd.
The registered company address is: 152 Beach Road, #21-01/04 Gateway East, Singapore 189721, Singapore

Foreword by Ashish Garg

Since the 1800s, industrial, economic and population growth has resulted in the incessant use of fossil fuels that has led the world to the brink of a climate disaster. This is largely driven by the emission of greenhouse gases due to the burning of fossil fuels which have caused global warming and pollution of natural ecosystems, resulting in many sustainability challenges. For mankind to grow in a sustainable manner and secure a healthy future, it is vital to limit the global warming below 1.5 °C which, otherwise, is likely to result in dire environmental consequences for the planet and its habitants in the coming decades. Realizing the gravity of the matter, scientists and policymakers have come up with seventeen sustainable development goals, also endorsed by the United Nations. Perhaps the most important of these goals is deriving energy from non-polluting sources which requires us to make a transition from fossil fuel-based energy to clean energy. This necessitates the timely development and deployment of cost-effective and locally adaptable cleaner energy generation methods using renewable sources, energy storage methods to overcome the intermittency of renewable energy, carbon capture utilization and storage technologies, improved energy efficiencies, development of cleaner fuels such as hydrogen and developing policy frameworks to expedite this transition, ideally towards NetZero.

This issue of directions showcases IIT Kanpur's research innovations in developing above sustainable energy technologies, encompassing materials, devices, processes and systems as well as policy frameworks for sustainability. The timely development of these is crucial to meeting our clean energy goals.

I am confident that this issue will serve as an excellent resource for academicians, researchers, students and postdocs, clean energy professionals and policymakers who are all working towards a sustainable energy transition and building a cleaner and greener world.

Ashish Garg
Professor and Head, Department
of Sustainable Energy Engineering
Indian Institute of Technology Kanpur
Kanpur, Uttar Pradesh, India

Foreword by Dr. Sachchida Nand Tripathi

Sustainability has emerged as one of the defining challenges of our times, requiring an interdisciplinary approach to bring advances in science, technology and policy. At IIT Kanpur, we take pride in contributing to this global endeavour through pathb-reaking research across diverse fields, including renewable energy, material science, sustainable industrial practices and environmental engineering. The vision of Kotak School of Sustainability is to foster collaborations and innovations aimed at addressing critical issues related to climate change and resource management. It gives me immense pleasure to present this insightful compilation, *Materials, Devices, and Systems for Sustainability*, as part of the *Directions* series of IIT Kanpur. This book stands as a testament to the remarkable synergy between innovative science and the forward-thinking policy of various departments of the institute.

The authors, through their rigorous exploration of cutting-edge advancements, provide an invaluable resource for researchers, industry leaders and policymakers committed to driving impactful change. From innovative strategies to enhance solar energy systems and clean hydrogen production to frameworks for grid resiliency and zero-carbon industrial transitions, the topics covered in this book reflect the institute's commitment to shaping a sustainable and greener future. This compilation highlights the transformative potential of novel materials, advanced devices and resilient systems.

It is my sincere hope that this compilation will inspire collaboration and catalyse meaningful advancements, both in research and practice, as we collectively work towards achieving global sustainability goals.

Dr. Sachchida Nand Tripathi
Dean, Kotak School of Sustainability
Indian Institute of Technology Kanpur
Kanpur, Uttar Pradesh, India

Series Editor's Preface

I feel elated about this special issue of Directions which has focused on research and development in the areas of sustainability and enabling technologies. At this point, I would also like to explain the genesis of the current issue.

Global warming and the consequent change in weather patterns require urgent collective action by policymakers, academia, researchers, businesses, other organizations and the society at large. This includes mitigation and adaptation measures while building resilience to climate change. In this regard, it is particularly important to facilitate research and innovation to help in developing and scaling up new, clean and renewable energy technologies, establishing new businesses, developing new skills, formulating supportive policies, devising the needed financial instruments, furthering education and outreach followed by sustained frugal practices and developing sustainable models for development. Also, although the cause is global, the solutions and approaches must be contextual and local and rather simple to implement by general public.

Following this broad outline, our editors have successfully identified a gamut of research works in the areas such as Energy Policy/Net Zero, Power sector reforms and regulations, Batteries, Hydrogen energy-related areas, Solar thermal engineering and Solar photovoltaics.

I am thankful to the authors for their spontaneous participation to contribute these articles and make timely revisions based on editors' notes. I am also thankful to the editors and the publication team for raising the quality of the publication. It is worthwhile to note that this is the fifth issue of Directions which is published by Springer. Based on my experience with the earlier issues, I expect to achieve a wider circulation for this issue as well, and surely the quality of the two-stage editing process will be appreciated by the readers. Any candid feedback is very much appreciated.

Prof. Tarun Gupta
Dean Research and Development
Indian Institute of Technology Kanpur
Kanpur, Uttar Pradesh, India

Preface

The 17 Sustainable Development Goals (SDGs) developed by UN calls for urgent action to achieve environmental, social and economic sustainability by 2030. At the centre of these interconnected goals lies the challenge of tackling climate change. Renewable and sustainable energy sources are key in addressing this challenge, especially as global CO_2 emissions from fossil fuels reach a record-high 37.4 billion tonnes in 2024, accounting for 75% of global greenhouse gas emissions. India's contribution stands at 4.6%, with emissions expected to rise due to the strong growth of the Indian economy which drives higher energy demand and accelerated infrastructure development. This highlights the critical need for adopting and utilizing emergent energy technologies to mitigate harmful emissions.

This edition of *Directions* is dedicated to showcasing the innovative work emerging from IIT Kanpur in materials, device developments, system technologies and policy frameworks for sustainable development, making it extremely relevant and timely. To achieve India's national goal of net-zero emissions by 2070, as set forth by the Government of India, the development and deployment of key energy technologies such as solar energy, hydrogen, batteries, carbon capture, biofuels and integrated systems will be pivotal. Along with these, techno-economic policy development will enable the government to make informed and balanced decisions, improving quality of life and driving economic growth. These aspects are included in this edited volume, covered across eleven chapters that showcase ongoing research and development, ranging from renewable energy generation, storage and distribution to policy analysis and implementation strategies.

In addition to highlighting recent work conducted by the authors in advancing the understanding, applicability and integration of various energy systems, this volume also provides policy and regulatory frameworks for their adoption, design and deployment so as to facilitate the realization of India's ambitious decarbonization goals. The chapters themselves have been carefully compiled by experts in their respective fields having wide experience not only in the technical domains but also in bridging insights from the perspective of both academia and industry. By presenting these cutting-edge findings to a broader audience, the editors of this volume endeavour to amplify the impact of these advancements on both national and global sustainability goals.

Kanpur, India

Sudarshan Narayanan
Laltu Chandra
Deepika Jhajhria

Contents

Strain–Relaxation Strategies for Perovskite Solar Cells

Sudhir Ranjan, Saurabh Srivastava, Shambhavi Rai, Anand Singh, Ashish Garg, and Raju Kumar Gupta

1 Introduction

In the past decade, organic–inorganic lead halide-based perovskites have emerged as the most promising light absorbers for next-generation energy-efficient photovoltaic devices. Their excellent optoelectronic properties have pushed single-junction perovskite solar cells (PSCs) to demonstrate high power conversion efficiencies (PCE), exceeding 26% (best at 26.1%), making them compete with silicon solar cells [1]. However, the PCEs of PSCs in single-junction architecture is still below the theoretical Shockley Queisser (SQ) limit of 33.7%, mainly due to high open-circuit voltage (V_{oc}) deficit owing to interfacial and bulk defects. Further, the devices exhibit long-term stability challenges, hindering their practical applications. Soft nature of perovskites leads to the generation of residual strain during crystallisation and growth of the perovskite thin films [2–5]. Several factors responsible for the origin of the strain include lattice parameter as well as coefficient of thermal expansion mismatch between the perovskite film and the substrate, nature of interface between the perovskite and the charge transport layers, atomic displacements and external bias. Usually, strain lowers the activation energy required for the defect formation, thereby facilitating the non-radiative recombination [4]. This negatively impacts the V_{oc} in the device thereby limiting the photovoltaic performance and long-term stability.

Strain engineering has gained momentum in recent times for minimising interfacial and bulk defects to minimise energy losses at the charge transport layer/perovskite interface and within the bulk of the perovskite. This portrays strain engineering as a powerful strategy to improve the photovoltaic performance and long-term stability of the devices for a wide range of applications. The fragile nature of perovskites further makes strain engineering an effective approach to relieve strain in the perovskite thin films. This motivates us to critically understand the origins of strain at different scales and develop effective engineering strategies for efficient management of strain in perovskite thin films for enhanced photovoltaic performance and long-term stability. This article briefly outlines the fundamentals of the strain in perovskite thin films, different methods for characterising the strain and its impact on the optoelectronic properties and stability of perovskite thin films and devices. We critically examine recent strain engineering strategies that have been employed to control strain in the perovskite thin films to improve device performance and long-term operational stability. We believe that developing a robust understanding of strain in the perovskite and their effective management would

S. Ranjan · R. K. Gupta (✉)
Department of Chemical Engineering, Indian Institute of Technology Kanpur, Kanpur, India
e-mail: guptark@iitk.ac.in

S. Srivastava
Department of Materials Science and Engineering, Indian Institute of Technology Kanpur, Kanpur, India

S. Rai · A. Garg (✉) · R. K. Gupta
Department of Sustainable Energy and Engineering, Indian Institute of Technology Kanpur, Kanpur, India
e-mail: ashishg@iitk.ac.in

A. Singh (✉)
Department of Chemistry, Indian Institute of Technology Kanpur, Kanpur, India
e-mail: anands@iitk.ac.in

A. Singh · A. Garg · R. K. Gupta
Chandrakanta Kesavan Centre for Energy Policy and Climatic Solutions, Indian Institute of Technology Kanpur, Kanpur, India

Kotak School of Sustainability, Indian Institute of Technology Kanpur, Kanpur, India

R. K. Gupta
Centre for Environmental Science and Engineering, Indian Institute of Technology Kanpur, Kanpur, India

S. Narayanan et al. (eds.), *Materials, Devices and Systems for Sustainability*, IITK Directions 8,
https://doi.org/10.1007/978-981-96-7295-0_1

further lead to future breakthroughs in device performance and long-term operation stabilities of PSCs. This would help develop multijunction PSCs (perovskite/perovskite tandem and perovskite/Si) where multiple layers contribute to stress and strain components.

2 Origin of Strain in Halide Perovskites

Strain arises in halide perovskites from a diverse range of factors. Strain inevitably emanates in the perovskite films during solution processing of the perovskite precursor solution on the substrates. The mismatch between the coefficient of thermal expansion (CTE) of perovskite film and the substrate induces strain during the cooling process of the perovskite film. When film is deposited on a substrate having lower CTE than that of the perovskite, the substrate constrains the perovskite's contraction along the in-plane directions, resulting in an in-plane tensile strain. Conversely, the deposition of a film on a substrate with a higher CTE than the perovskite results in a compressive strain. The cooling process followed by annealing induces strain, typically in the form of tensile strain, within the plane of the film [6]. In addition, the fundamental nucleation and crystal development processes advance the perovskite crystallisation. As the solvent evaporates during spin-coating, nuclei are created at interfaces or inside the bulk of precursor film. The viscoelastic creep of the precursor solution and the chemical coordination between the solute and solvents control the bonding between the perovskite and the substrate, which limits the crystal formation. The main challenge is the inability to control nucleation and crystal development, which results in residual microstrain in perovskite films. These films are characterised by a broad distribution of grain sizes as well as variations in grain-to-grain orientation [5, 7].

Moreover, depending on whether the substrate's d-spacing is greater or smaller than the perovskite layer's, the lattice mismatch between the two can result in either compressive or tensile strain. Non-uniform interfacial distortion results from the perovskite crystals' tendency to stack randomly on the underlying crystal in cases of significant lattice mismatch [8, 9]. In addition, hydrostatic pressure and mechanical bending of the film can be used to introduce [10] compressive and tensile strains to films [11]. In halide perovskites, additional factors such as atomic misfits, phase transitions, light/bias stimulation and grain boundaries are associated with local strain, specifically microstrain; however, all of the previously listed factors often impact strains over the entire film [12, 13]. Additionally, when larger cations and anions occupy positions in the perovskite lattice, tilting of the PbI_6 octahedral results in elevated local lattice distortion [14]. In the same way, phase transitions can also increase the microstrain because they usually result in less symmetry in the crystal structure than in the cubic perovskite structure.

Ion migration and/or phase segregation aided by light can also enhance the microstrain in halide perovskites due to the interaction of photoexcited carriers with the soft, ionic lattice of the perovskite [15, 16]. Another source of strain in these materials comes from the perovskite film's grain boundaries. The resulting strain is dependent on particular interfacial conditions, like orientation mismatch at boundaries, which vary significantly within thin films [17]. Due to the high degree of lattice distortion and non-stoichiometric chemical distribution at the grain boundaries, large strain fields are produced [4, 18]. Sub-grain lattice orientation disorder and twin boundaries, in addition to grain boundaries, can also function as sources of strain in halide perovskite films [17, 19] (Fig. 1).

3 Strain Controls Properties of Perovskite Thin Films and Crystals

The strain has a direct impact on the perovskite structure, thereby influencing both the optoelectronic properties and the stability of the material, whose extent is governed by the material [12]. From a device optimisation perspective, it is crucial to minimise the non-radiative recombination in the perovskite films to reduce the energy loss. The presence of deep traps in the halide perovskites poses a significant challenge to device performance. In the perovskite films, substantial variations in intragrain crystallographic orientations, coupled with corresponding local strains, are associated with elevated trap densities, [17, 20, 21] which in turn, give rise to non-radiative recombination. This implies that effectively managing the local strain is pivotal in minimising the non-radiative recombination in these materials [12].

Strain is essential for altering the bond length (Pb–X) and bond angles (Pb–X–Pb) in the band structure of halide perovskites. Modifications in orbital overlaps result in variations in band dispersion [22]. Typically, the bandgap increases as compressive and tensile strains increase [23]. It has been observed that when compressive strain increases, carrier mobility increases as well [24]. Moreover, strain affects ion migration's activation energy, which has been found to rise with compressive strain and fall with tensile strain. It controls the thermodynamics of defect generation, which has a major impact on the stability of halide perovskites. Also DFT calculations indicate that when tensile strain increases, the defect production energy of halide vacancies reduces [25].

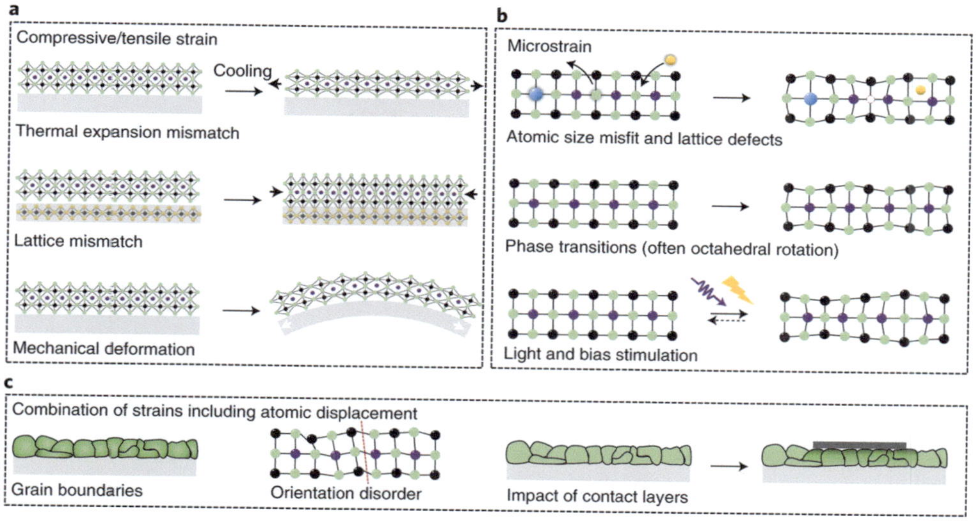

Fig. 1 Different origins of strain in halide perovskites: **a** origin of compressive and tensile strain, **b** microstrain, **c** combination of strains in halide perovskites caused by both shear and rotational components. A-site cations (MA, FA, Cs) are represented by blue and purple spheres, whereas metal cations (Pb, Sn) and halide anions (I, Br, Cl) are represented by black and green spheres, respectively. The substrate atoms are shown by orange spheres in (a), vacancy and extrinsic dopant are shown by a red dashed circle and yellow sphere, respectively, in (b), and the twin boundary is represented by a red dashed line. Reproduced with permission from Ref. [12]. Copyright 2021, Springer Nature

4 Strain Regulation Strategies in Perovskites

Minimisation of the unwanted energy losses originating from the interfacial and bulk defects in the perovskite layer have led to development of energy-efficient devices with long-term operational stability. Recently, strain engineering has been utilised to minimise the formation of undesirable defects and subsequently mitigate non-radiative recombination for fabricating highly efficient and long-term stable devices [14, 24, 26–28]. Numerous strategies, including modifying the annealing procedure, adjusting the chemical composition of the film and investigating the role of interfaces, have been researched for improving control over strain in perovskite films. These techniques fall into two categories based on the different sources of strain: (i) regulation of strain induced by local conditions; (ii) regulation of strain induced by external conditions (Fig. 2).

4.1 Regulation of Local Strain

4.1.1 Composition Tailoring

In addition to improving the bandgap through halide substitution on X site, compositional tuning of perovskites can also be used to improve thermal stability through Cs and FA substitution on the A site. This phenomenon related to local strain relaxation induced by substitution has been observed recently. A few studies have suggested that the addition of ions with radii smaller than FA (253 pm), such as MA (217 pm) and Cs (167 pm) at the A site and/or Br (196 pm) and Cl (181 pm) at the X site, could reduce the microstrain in $FAPbI_3$-based systems [3, 14, 31]. This could be the result of a potential strain relaxation mechanism whereby these additives decrease the B–X–B bond angles and B–X bond lengths, as determined by density functional theory calculations [3]. An alternative method involves integrating bigger cations, like PEA, to create 2D or 2D/3D perovskites. This has resulted in improved device efficiency compared to the 3D-only control samples, with a reduction of diffraction peak width and increase in the Bragg peak area, which are generally signs of decreased microstrain [32] (Fig. 3).

4.1.2 Mitigating Inhomogeneity in Perovskite Films

Chen et al. [23] controlled the perovskite annealing process and proposed a method of flipped annealing of perovskite films to modify the in-plane strain gradient in response to the local lattice strain caused by the inhomogeneity of mixed halide perovskites. To eliminate the temperature gradient that results in compositional inhomogeneity during the fabrication of perovskite films, authors specifically replaced the conventionally annealed perovskite film with a flipped

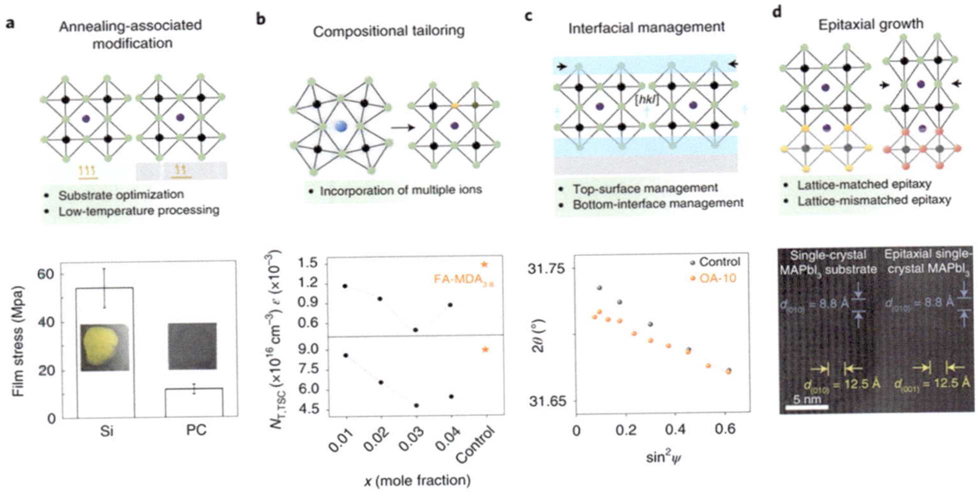

Fig. 2 **a** Modification associated to annealing: schematic diagram of preventing strain generation due to annealing (on top) and the bottom one shows the stress measured for $MAPbI_3$ coated on silicon (Si) and polycarbonate (PC) substrates at an annealing temperature 100 °C. Images of the perovskite deposited on the two substrates after 45 h of dry heat ageing (at 85 °C and 25% relative humidity) are shown in the insets. It is only on the Si substrate, where there is a larger mismatch of thermal expansion, that the perovskite breaks down into the yellow phase [10]. Copyright 2018, Wiley–VCH **b** composition tailoring. Top: a schematic illustrating the incorporation of several ions.Bottom: microstrain calculated in samples of mesoporous $TiO_2/(FAPbI_3)_{1-x}(MC)_x$-coated glass/fluorine-doped tin oxide (FTO) (upper plot) and trap density NT, TSC (lower plot), where the orange star indicates the control perovskite $((FAPbI_3)_{0.972}(MDACl_2)_{0.038})$ and MC is an equimolar mixture of MDA^{2+} and Cs^+. [12] Copyright 2021, Springer Nature. **c** Interfacial management. Top: An example of managing the perovskite interface. Bottom: 2θ versus $\sin^2\psi$ linear fit for control and n-octylammonium modified (OA-10) perovskite films. A lower slope in perovskite films indicates a lower tensile stress [29]. Copyright 2019, Wiley–VCH Verlag. **d** Epitaxial growth. Top: a diagram illustrating epitaxial growth. Bottom: a high-resolution transmission electron microscopy (HRTEM) image displaying the $MAPbI_3$ substrate's epitaxial single-crystal interfacial area [30]. Copyright 2020, Springer Nature

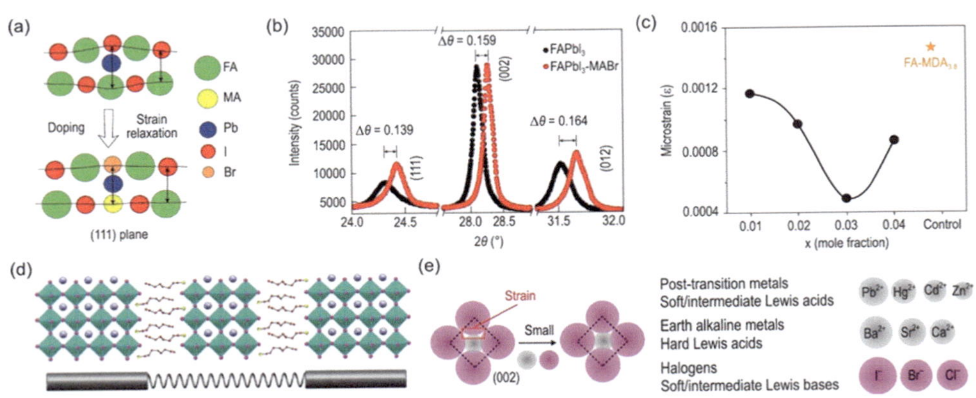

Fig. 3 **a** A diagram of the strain in the $FAPbI_3$ plane (111) and the strain relaxation following MABr alloying. **b** $FAPbI_3$ and $FAPbI_3$–MABr XRD patterns [14]. Copyright 2016, The American Chemical Society. **c** Perovskite with a structure of FTO/mp-TiO2/perovskite with calculated strain [33]. Copyright 2021, China Science Publishing and Media Ltd. **d** A schematic illustration of stress relaxation using stiff and soft structural components [29]. Copyright 2019, Wiley–VCH Verlag. **e** Diagram showing how strain is reduced by adding tiny B/X-site ions [3]. Copyright 2018, Springer Nature

annealing method, resulting in the release of tensile strain and introduction of a compressive strain, as shown by the GIXRD data (Fig. 4a and b). Tensile-strained devices only showed an average efficiency of ~18.7%, while the optimised strain-free PSC eventually achieved a certified PCE of 20.7% with an average of 19.8% (Fig. 4).

4.1.3 Lattice Expansion-Induced Local Strain Regulation

It has been shown that light-induced structural dynamics are essential to stability and physical characteristics of perovskites, as well as performances of corresponding optoelectronic devices. Apart from the commonly noted effects

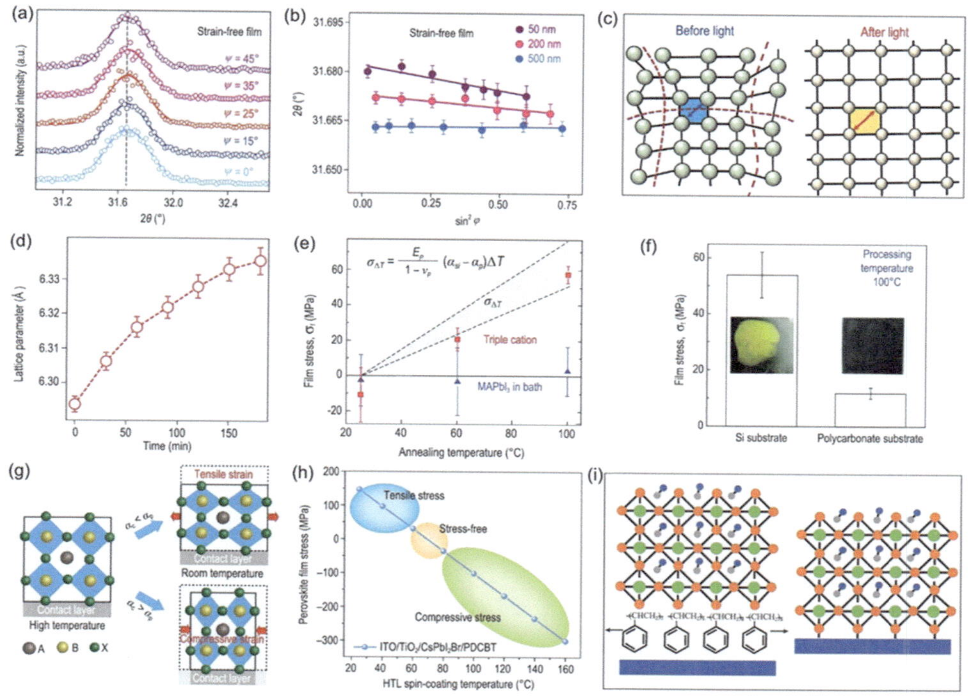

Fig. 4 **a** GIXRD spectra for the strain-free film at various tilt angles and at a depth of 50 nm. **b** The strain distribution for the strain-free film at 50, 200 and 500 nm depths [34]. Copyright 2019, Springer Nature. **c** Schematic showing how the structure changes both before and after illumination (lattice expansion versus local distortion). **d** The dependency lattice constant on illumination time [33]. Copyright 2021, China Science Publishing and Media Ltd. **e** Stress in $MAPbI_3$ and CsMAFA films formed at varying temperatures, measured and calculated. **f** Measured stress in $MAPbI_3$ films formed on Si and polycarbonate at 100 °C; pictures of $MAPbI_3$ on Si and polycarbonate taken 45 h later at 85 °C and 25% relative humidity are shown in the insets [10]. Copyright 2018, Wiley–VCH Verlag. **g** Schematic showing how compressive and tensile strains form. **h** The PDCBT spin-coating temperature-dependent stress calculated for perovskites inside ITO/TiO_2/perovskite/PDCBT structures. [25]. Copyright 2020, Springer Nature. **i** Schematic showing the mechanism of the PS buffer layer that releases the stress in perovskites [35]. Copyright 2019, Wiley–VCH Verlag

of light on ion migration and halide segregation, Mohite et al. have reported that illumination can also cause a relaxation of local strain in perovskites [36]. They found that a typical 1-sun (100 mW cm^{-2}) light source used continuously causes large and consistent lattice expansion in perovskite films. It is believed that photogenerated electron–hole pairs weaken covalent bonds, cause the Pb–I–Pb bonds to elongate or become less distorted, and ultimately cause the lattice to expand, resulting in the relaxation of local strain in perovskite. Such expansions are particularly prominent in mixed-cation perovskites, which are typically under strain because of the lattice's distortion from the cations of varying sizes.

4.2 Regulating External Condition-Induced Strain

The strain caused by external conditions is generally because of two types of mismatches between substrate and perovskite, namely, lattice mismatch and the coefficient of thermal expansion mismatch.

4.2.1 Regulation of Strain via Annealing

Among the above-discussed cases, primary cause of strain in perovskite solar cells is the thermal expansion mismatch. Several strain modulation techniques have been recently reported in relation to the stress and thermal expansion mismatch correlation. These methods fall into one of two categories: (i) reducing the temperature at which perovskite films form to lower temperature gradient when perovskite films cool from the annealing temperature to room temperature; [34] and (ii) utilising contacting layers with perovskites that have similar linear thermal expansion coefficient to lower the thermal expansion difference between the perovskite and the substrate [36]. Temperature changes cause stress to build up as larger stress values are obtained at higher annealing temperatures. On the other hand, there would be a significant reduction in stress if a perovskite film is deposited at a low temperature, even at room temperature [10]. A different low-temperature method that uses room-temperature amine-assisted crystallisation has allowed photovoltaic devices to achieve a high PCE of 23.1% [36]. While these reports indicate the possibility of modifying the deposition process concerning annealing, more innovation is needed to prevent significant residual strains in films. Insights into any potential hidden roles played by solvent molecules may also be investigated by comparing solution-processed films with films obtained by thermal evaporation, where no annealing is required to form a photoactive black perovskite phase [37, 38]. Additionally, during annealing, there is a vertical temperature gradient across device thickness, which can create gradients along this direction for chemicals and strain. However, it is challenging to characterise these features without the use of multidimensional and depth-dependent techniques.

4.2.2 Interfacial Engineering Regulates Lattice Mismatch-Induced Strain

A well-known method of material growth, epitaxy involves thoughtfully choosing the substrates that are lattice-mismatched/matched to introduce controlled levels of strains into materials. Epitaxial growth of α-$FAPbI_3$ has been accomplished recently through the modification of a $MAPbCl_xBr_{3-x}$ single-crystal substrate [24]. Further modifications in the composition also made it possible to apply varying degrees of strain to the epitaxial α-$FAPbI_3$. It was found that lower compressive strain significantly improved the phase stability. In another study, the use of lattice-matched $MAPbI_3$ substrates allowed for the solution-phase epitaxial growth of $MAPbI_3$, resulting in a remarkably large area (>25 cm^2) and the absence of dislocations [30]. It may be beneficial to grow halide perovskites heteroepitaxial directly on top of the charge transport layers. However, doing so will further require growing the current library of halide contact layers. The tendency of phase impurities to epitaxially grow alongside the perovskite phases is another intriguing feature of epitaxial growth [39]. It is crucial to reduce phase impurities, particularly those detrimental to stability and performance. Hence, more investigation is required to comprehend how epitaxial growth stabilises halide perovskites. However, the use of epitaxial techniques carries a risk of negating the main benefit of halide perovskites, which is their inexpensive and scalable processability, in favour of more stable and efficient III–V or silicon-based alternatives.

5 Summary

Developing a fundamental understanding of strain in the perovskite thin film/crystals and their effective management would further lead to achieving breakthroughs in device performance and long-term operation stabilities of PSCs in the future. This would help in developing multijunction PSCs (perovskite/perovskite tandem and perovskite/Si) where multiple layers contribute to both stress and strain components. In the coming years, as the photovoltaic community thrives towards achieving high-performing devices complemented with long-term operational stability, strain manipulation will serve as a critical factor for device optimisation and commercialisation.

Acknowledgements Authors acknowledge the financial support from the Department of Science and Technology (DST), India, through Grant DST/TMD/CERI/C140(G) under the Clean Energy Research Initiative; and UKRI Global Challenge Research Fund through project SUNRISE (EP/P032591/1).

Notes The authors declare no competing financial interest.

References

1. Park J et al (2023) Controlled growth of perovskite layers with volatile alkylammonium chlorides. Nature 616(7958):724–730
2. Kim G et al (2020) Impact of strain relaxation on performance of α-formamidinium lead iodide perovskite solar cells. Science 370(6512):108–112
3. Saidaminov MI et al (2018) Suppression of atomic vacancies via incorporation of isovalent small ions to increase the stability of halide perovskite solar cells in ambient air. Nat Energy 3(8):648–654
4. Jones TW et al (2019) Lattice strain causes non-radiative losses in halide perovskites. Energy Environ Sci 12(2):596–606
5. Ranjan R et al (2021) Enhanced thermal and moisture stability via dual additives approach in methylammonium lead iodide based planar perovskite solar cells. Sol Energy 225:200–210
6. Zhao J, et al (2017) Strained hybrid perovskite thin films and their impact on the intrinsic stability of perovskite solar cells. Sci Adv 3(11):eaao5616
7. Liu C, Cheng Y-B, Ge Z (2020) Understanding of perovskite crystal growth and film formation in scalable deposition processes. Chem Soc Rev 49(6):1653–1687
8. Zhang B et al (2020) NiO/perovskite heterojunction contact engineering for highly efficient and stable perovskite solar cells. Adv Sci 7(11):1903044
9. Zhang CC et al (2020) Perovskite films with reduced interfacial strains via a molecular-level flexible interlayer for photovoltaic application. Adv Mater 32(38):2001479
10. Rolston N et al (2018) Engineering stress in perovskite solar cells to improve stability. Adv Energy Mater 8(29):1802139
11. Jiao Y et al (2021) Strain engineering of metal halide perovskites on coupling anisotropic behaviors. Adv Func Mater 31(4):2006243
12. Liu D et al (2021) Strain analysis and engineering in halide perovskite photovoltaics. Nat Mater 20(10):1337–1346
13. Ranjan S et al (2022) Low-temperature microwave processed TiO2 as an electron transport layer for enhanced performance and atmospheric stability in planar perovskite solar cells. ACS Appl Energy Mater 5(3):2679–2696
14. Zheng X et al (2016) Improved phase stability of formamidinium lead triiodide perovskite by strain relaxation. ACS Energy Lett 1(5):1014–1020
15. Bischak CG et al (2017) Origin of reversible photoinduced phase separation in hybrid perovskites. Nano Lett 17(2):1028–1033

16. Muscarella LA et al (2020) Lattice compression increases the activation barrier for phase segregation in mixed-halide perovskites. ACS Energy Lett 5(10):3152–3158
17. Jariwala S et al (2019) Local crystal misorientation influences non-radiative recombination in halide perovskites. Joule 3(12):3048–3060
18. Hÿtch M, Putaux J-L, Thibault J (2006) Stress and strain around grain-boundary dislocations measured by high-resolution electron microscopy. Phil Mag 86(29–31):4641–4656
19. Xiao X et al (2020) Benign ferroelastic twin boundaries in halide perovskites for charge carrier transport and recombination. Nat Commun 11(1):2215
20. Srivastava S et al (2023) Advanced spectroscopic techniques for characterizing defects in perovskite solar cells. Communications Materials 4(1):52
21. Ranjan R et al (2018) Effect of tantalum doping in a TiO 2 compact layer on the performance of planar spiro-OMeTAD free perovskite solar cells. J Mater Chem A 6(3):1037–1047
22. Ghosh D et al (2019) Putting the squeeze on lead iodide perovskites: pressure-induced effects to tune their structural and optoelectronic behavior. Chem Mater 31(11):4063–4071
23. Zhu C et al (2019) Strain engineering in perovskite solar cells and its impacts on carrier dynamics. Nat Commun 10(1):815
24. Chen Y et al (2020) Strain engineering and epitaxial stabilization of halide perovskites. Nature 577(7789):209–215
25. Xue D-J et al (2020) Regulating strain in perovskite thin films through charge-transport layers. Nat Commun 11(1):1514
26. Hu M et al (2021) Regulating the surface passivation and residual strain in pure tin perovskite films. ACS Energy Lett 6(10):3555–3562
27. Kapil G et al (2019) Strain relaxation and light management in tin–lead perovskite solar cells to achieve high efficiencies. ACS Energy Lett 4(8):1991–1998
28. Shai X, et al (2023) Origins, impacts, and mitigation strategies of strain in efficient and stable perovskite solar cells. Small Sci 2300014
29. Wang H et al (2019) Interfacial residual stress relaxation in perovskite solar cells with improved stability. Adv Mater 31(48):1904408
30. Lei Y et al (2020) A fabrication process for flexible single-crystal perovskite devices. Nature 583(7818):790–795
31. Saliba M et al (2016) Incorporation of rubidium cations into perovskite solar cells improves photovoltaic performance. Science 354(6309):206–209
32. Lee J-W et al (2018) 2D perovskite stabilized phase-pure formamidinium perovskite solar cells. Nat Commun 9(1):3021
33. Wu J, et al (2021) Strain in perovskite solar cells: origins, impacts and regulation. Natl Sci Rev 8(8):nwab047
34. Brennan MC et al (2017) Light-induced anion phase segregation in mixed halide perovskites. ACS Energy Lett 3(1):204–213
35. Wu J et al (2019) A simple way to simultaneously release the interface stress and realize the inner encapsulation for highly efficient and stable perovskite solar cells. Adv Func Mater 29(49):1905336
36. Wang K et al (2020) Isothermally crystallized perovskites at room-temperature. Energy Environ Sci 13(10):3412–3422
37. Dewi HA et al (2021) Excellent intrinsic long-term thermal stability of co-evaporated MAPbI3 solar cells at 85° C. Adv Func Mater 31(22):2100557
38. Chiang Y-H, Anaya M, Stranks SD (2020) Multisource vacuum deposition of methylammonium-free perovskite solar cells. ACS Energy Lett 5(8):2498–2504
39. Rothmann MU, et al (2020) Atomic-scale microstructure of metal halide perovskite. Science 370(6516):eabb5940

Challenges in Solid-State Electrolyte-Based Batteries and the Roadmap to Commercialization

Mohammed Saquib Khan, M. Dinachandra Singh, Pragati Singh, Kanwar S. Nalwa, and Sudarshan Narayanan

1 Introduction

Since their introduction by Sony in 1990, lithium-ion (Li-ion) batteries have gained substantial market share, owing to benefits, for instance, superior energy density and prolonged cycle life [1]. Widely utilized in various applications, including portable electronic gadgets, hybrid and electric vehicles, and household power backups, etc. Li-ion batteries continue to dominate the market.

However, because they use extremely flammable organic electrolytes with low thermal stability and a low flame point, typical Li-ion batteries present serious safety risks. This makes them susceptible to fire accidents and explosions under improper usage conditions [2]. To comprehensively address these safety challenges, a shift away from highly flammable organic liquid electrolytes is imperative. All-solid-state batteries (ASSBs) emerge as a compelling alternative, replacing organic liquid electrolytes with inorganic solid electrolytes recognized for their extraordinary thermal stability [3]. Beyond enhanced safety, ASSBs offer several advantages, including simplified packaging owing to the nonexistence of a liquefied electrolyte, reduced dead weight, and increased energy density. Inorganic solid electrolytes also exhibit superior electrochemical stability, enabling compatibility with advanced cathode materials and further enhancing energy density. Additionally, ASSBs demonstrate exceptional mechanical qualities.

While Li-ion batteries utilizing liquid organic electrolytes display favorable functioning attributes such as good electrical conductivity and interfacial contact, they come with drawbacks, including lower ion selectivity, inadequate constancy, and intensified volatility as a result of igneous organic liquids, posing significant safety risks [4]. The surge in interest in ASSBs in recent years is notable, driven by their inherent safety advantages, prolonged cycle stability, and enhanced thermal stability [5].

Solid-state electrolytes (SSEs), a vital component in ASSBs, fall into three groups: organic polymer SSE [6], inorganic SSE [7], and organic–inorganic composite SSE [8]. To achieve higher energy density than traditional liquid batteries, the development of solid electrolytes capable of accommodating higher voltages becomes imperative [9].

The ASSBs stand out from traditional batteries by substituting the organic liquefied electrolyte and separator by an inorganic solid electrolyte [10]. Despite this distinction, the operational principles remain consistent. During the charging phase, Li-ions deintercalate from the cathode, traverse the electrolyte and electrolyte–electrode interfaces, and reach the anode. Simultaneously, electrons migrate from the cathode to the anode via peripheral circuit. The discharge process involves the reversible reversal of Li-ion and electron movement. Solid electrolytes, similar to liquid electrolytes and separators, serve the fundamental functions of permitting only Li-ion exchange between anode and cathode, and preventing short circuits. Thus, certain crucial prerequisites for solid electrolytes are imperative, including ionic conductivity exceeding 1×10^{-4} S cm^{-1} at room temperature, high electronic resistivity with an elevated ionic transfer, and broad electrochemical potential window for high-voltage operation [11].

M. S. Khan · M. D. Singh · P. Singh · K. S. Nalwa · S. Narayanan (✉)
Department of Sustainable Energy Engineering, Indian Institute of Technology Kanpur, Kanpur, India
e-mail: sudarshan@iitk.ac.in

S. Narayanan
Kotak School of Sustainability, Indian Institute of Technology Kanpur, Kanpur, India

S. Narayanan et al. (eds.), *Materials, Devices and Systems for Sustainability*, IITK Directions 8,
https://doi.org/10.1007/978-981-96-7295-0_2

2 Solid-State Electrolytes (SSEs)

2.1 Sulfide Solid Electrolytes

In comparison to their oxide-based equivalents, sulfide-based materials have better Li^+ ion conductivity at room temperature, which makes them more appropriate for use in batteries [12]. Additionally, sulfide-based electrolytes possess relative softness and deformability. The higher polarizability of sulfide-based electrolytes, in comparison to oxide-based ones, results in weaker bonding between Li^+ ions and the sulfide framework. This, in turn, leads to higher mobility for sulfide-based electrolytes compared to oxide-based ones.

Otto in 1996 reported a conductivity of 3.3×10^{-6} S cm^{-1} at 25 °C and 9.7×10^{-2} S cm^{-1} at the temperature of 350 °C for the Li_2O–Li_2Cl_2–Li_2SO_4–SiO_2–B_2O_3 glass system with a composition of 35:10:30:12.5:12.5 [13]. Cales et al. (1977) documented ionic conductivity of 1×10^{-3} S cm^{-1} at 300 °C for borate-based glassy electrolytes, specifically in systems such as B_2O_3–Li_2O–LiX (where X = Cl, Br, F, I) and Li_2O–Li_2SO_4-B_2O_3. These results served as a catalyst for additional investigations into novel sulfide-type electrolyte systems. Mercier et al. discovered that the conductivity of Li_2S–P_2S_5–LiI is ~1×10^{-3} S cm^{-1} at room temperature [14]. Ribes and Pradel investigated Li_2S–M (M = SiS_2, B_2S_3, GeS_2, As_2S_3, P_2S_5) and $xLi_2S_{(1-x)}SiS_2$ ($x \leq 0.6$) glasses. In 1987, Kennedy presented findings on the synthesis through melt quenching and conductivity studies of Li_2S–SiS_2 with lithium halides (LiX, where X = I, and Br). [15]. Molecular dynamics simulation studies were performed by Rao and Seshasayee [16] on superionic sulfide glasses' ternary systems x(0.5Li_2S–0.5P_2S_5)–(1 − x)LiI) and (x(0.4$Li_2S_{0.6}P_2S_5$)–(1 − x)LiI. They attributed the notable ionic conductivity at room temperature in these systems to the occurrence of non-linking sulfur atoms surrounding dispersing lithium atoms [16]. The presence of iodine atoms contributed to a decrement in the glass-transition temperature (T_g) by plasticizing the structure, making it less rigid. In 1995, it was reported by Takada et al. that ASSBs with thin-film based cells featuring the Li/Li_3PO_4–Li_2S–SiS_2/$LiMO_2$ (M = Co, Ni) architecture exhibited capacities ranging from 80 to 90 mAh g^{-1} when cycled in the potential window of 2.0–3.8 V at 64 μA cm^{-2} [17]. Subsequently, researchers worldwide explored various glassy and nanocrystalline-type sulfide-based electrolytes.

Numerous research groups have investigated a variety of electrolyte materials, including Li–P–S composite glasses, argyrodites, glass-based ceramics, Li_6PS_5X (X = Halide), thio-LISICONs, and $Li_{11-x}M_{2-x}$ P_{1+x} S_{12} (M = Ge, Sn, and Si) [18, 19]. $L_6P_8S_5Cl$, $Li_7P_2S_8I$, and β-Li_3PS_4 have garnered extensive attention in research because of their outstanding conductivity and excellent mechanical properties, which facilitate the construction of ASSBs. Another recently developed material, $Li_{10}GeP_2S_{12}$ (LGPS), has demonstrated an impressive ionic conductivity of 10^{-2} S cm^{-1}, similar to those of electrolytes in liquid form [20]. In LGPS, ion migration occurs along a one-dimensional (1D) pathway created by $(Ge_{0.5}P_{0.5})S_4$ tetrahedra and LiS_6 octahedra. To mitigate the challenges arising from the low availability and high costs associated with germanium (Ge), researchers have suggested replacing Ge^{4+} with either silicon (Si^{4+}) or tin (Sn^{4+}). Furthermore, $Li_{9.54}Si_{1.74}P_{1.44}S_{11.7}Cl_{0.3}$ exhibits the highest conductivity reported at 2×10^{-2} S cm^{-1}. This superior conductivity is attributed to a three-dimensional (3D) conduction approach according to the 1D route observed in LGPS [21].

2.2 Garnet Structured Electrolyte

Garnet-type solid electrolytes (GSSEs) offer notable advantages, including robust stability with Li metal, conductivity, and a broad potential window. The paramount characteristic of garnet-type SSEs is their stability versus Li metal [22]. Murugan et al. in 2007 synthesized the highly conductive $Li_7La_3Zr_2O_{12}$ (LLZO) garnet-type SSE, demonstrating an impressive ionic conductivity of 3.1×10^{-4} S cm^{-1} at 30 °C temperature [23]. The material demonstrated excellent thermal stability in thermogravimetric measurements conducted from 20 to 900 °C, with no observed weight loss, while demonstrating an activation energy of ~0.32 eV [24].

In a notable study, Yang et al. reported on a molten lithium battery based on garnet-type SSE, showcasing stable cycling at 240 °C for 1 month with minimal decay in battery capacity. Moreover, the battery showed a noteworthy good power capability, sustaining a current density of up to 300 mA cm^{-2}. Furthermore, the conductivity of the garnet-type solid-state electrolyte (SSE) reached as high as 85 mS cm^{-1} at 240 °C. The ionic conductivity of solid electrolytes generally rises with temperature, following a typical Arrhenius-type behavior.

Despite the garnet-type SSEs exhibiting a relatively high ionic conductivity that fulfill the criteria for marketable batteries, it remains lower in comparison to the organic liquid electrolytes, typically around 10^{-2} Scm^{-1} at room temperature [25].

The garnet-structured SSEs have a structural composition represented by the general formula $A_3B_3C_2O_{12}$, where A, B, and C cations are involved in eight, four, and six coordinations, respectively, in crystallographically unique cation coordination environments. The 3D assembly in the $B_3C_2O_{12}$ structure is established by tetrahedral 24d-A sites, bridged by a single octahedron. The A-site, typically engaged by a bigger cation, constitutes a dodecahedral interspace in the structure [27].

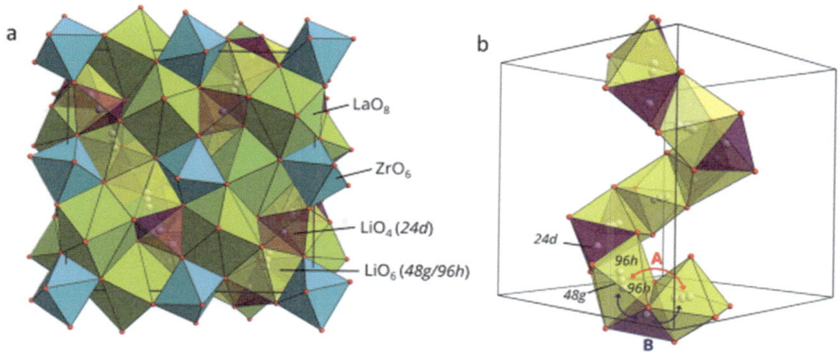

Fig. 1 **a** Illustrates the representative crystal structure of Lithium Lanthanum Zirconium Oxide (LLZO), **b** Wyckoff positions of the lithium-ion [26]

Figure 1a depicts crystal structure of LLZO, comprising octahedral ZrO_6 and dodecahedral LaO_8 elements [26]. Li atoms can be found at tetrahedral 24d sites, octahedral 48 g sites, or off-centered 96 h sites, which are examples of interstitial sites. Lithium ions are displaced within the octahedra from the central 48 g site to the off-centered 96 h site as a result of repulsive interactions between Li-ions. The tetrahedral cage at the 24d site is interconnected with four adjacent octahedral cages (Fig. 1b), creating a 3D network of pathways for lithium-ion migration within the framework of garnet [28]. Wang et al., Han et al., and Hayamizu et al. conducted studies utilizing various techniques, including nuclear magnetic resonance and neutron diffraction [18, 29]. These investigations aimed to explore the routes and behaviors of transport of Li-ion in garnet electrolytes. In a separate study, Thangadurai et al. introduced a novel class of fast Li-ion conductors in 2003, $Li_5La_3M_2O_{12}$ (M = Ta, Nb), featuring garnet structures, albeit exhibiting low conductivity (~10^{-6} S cm^{-1}). The substitution of Zr for the M-site element, as demonstrated by R. Wagner et al., led to LLZO with enhanced ionic conductivity, reaching ~3×10^{-4} S cm^{-1} [30]. It is noteworthy that garnet-type solid electrolytes manifest two crystal arrangements: tetragonal and cubic phases. The tetragonal phase is characterized by significantly lower ionic conductivity, attributed to Li^+ being fully occupied at specific sites within this structure.

Doping strategies incorporating super-valent cations such as Ta, Al, and Ga have been utilized to stabilize garnet-type SSEs in its cubic phase at room temperature. These approaches introduce vacancies at Li^+ sites and reduce ordering of local Li^+, ultimately resulting in an improvement in Li^+ conductivity. Various substitution cations, including Ba^{2+}, Ca^{2+}, Nb^{5+}, Ta^{5+}, Sr^{2+}, Y^{3+}, among others, have been explored, with Ta doping proving effective in enlarging Li1 and Li2 site sizes, facilitating Li^+ migration [31]. In addition to cation doping, anion doping such as the introduction of fluorine, has been effective in increasing ionic conductivity [32]. The microstructure of garnet SSEs, including factors like grain boundary, grain size, and specific density, influences ionic conductivity. Nanocrystalline materials with smaller grain size and larger grain boundary area exhibit enhanced ionic conductivity. Second-phase additives, exemplified by nano-MgO, can suppress grain growth during sintering, improving overall performance [33].

Aprotic synthetic processes, like the self-consolidation method, have been successful in achieving high relative density and bulk ionic conductivity. Engineering internal lithiation reservoirs and reducing sintering temperatures are proposed processing methods that contribute to lowering costs and promoting large-scale applications [34].

In conclusion, enhancing the ionic conductivity of garnet-type SSEs involves implementing effective strategies such as doping with high-valence ions, meticulous control of grain size, optimization of synthetic processes, and the incorporation of anion dopants.

2.3 NASICON (Sodium SuperIonic CONductor)

These substances are appropriate for use in solid-state lithium/sodium metal batteries owing to their great ionic conductivity and structural robustness. The NASICON-type $Na_{1+x}Zr_2P_{3-x}Si_xO_{12}$ structure was reported by Goodenough et al. [35] as early as 1976. NASICON's crystalline framework contains a 3D network structure with tunnels for Na^+ movement. The NASICON structure is derived by $NaM_2(PO_4)$ (M = Ge, Ti, or Zr) [36]. Its rhombohedral unit cell (space group $R\overline{3}c$) is made up of MO_6 octahedra and PO_4 tetrahedra that share corners (Fig. 2). Oxygen atoms are shared by three PO_4 tetrahedra and two MO_6 octahedra, resulting in the lantern-like units, which are combined to form a three-dimensional network. The structure has three distinct cation sites: Li1 (sixfold oxygen coordination), Li2 (tenfold oxygen coordination), and Li3 (between Li1 and Li2, fourfold oxygen coordination) [37].

Interestingly, when Li atoms replace the Na atoms, the material retains its NASICON structure and becomes a Li-ion conductor with good conductivity. These fast ion conductors (FICs) have the general formula $LiM_2(PO_4)_3$, which contains

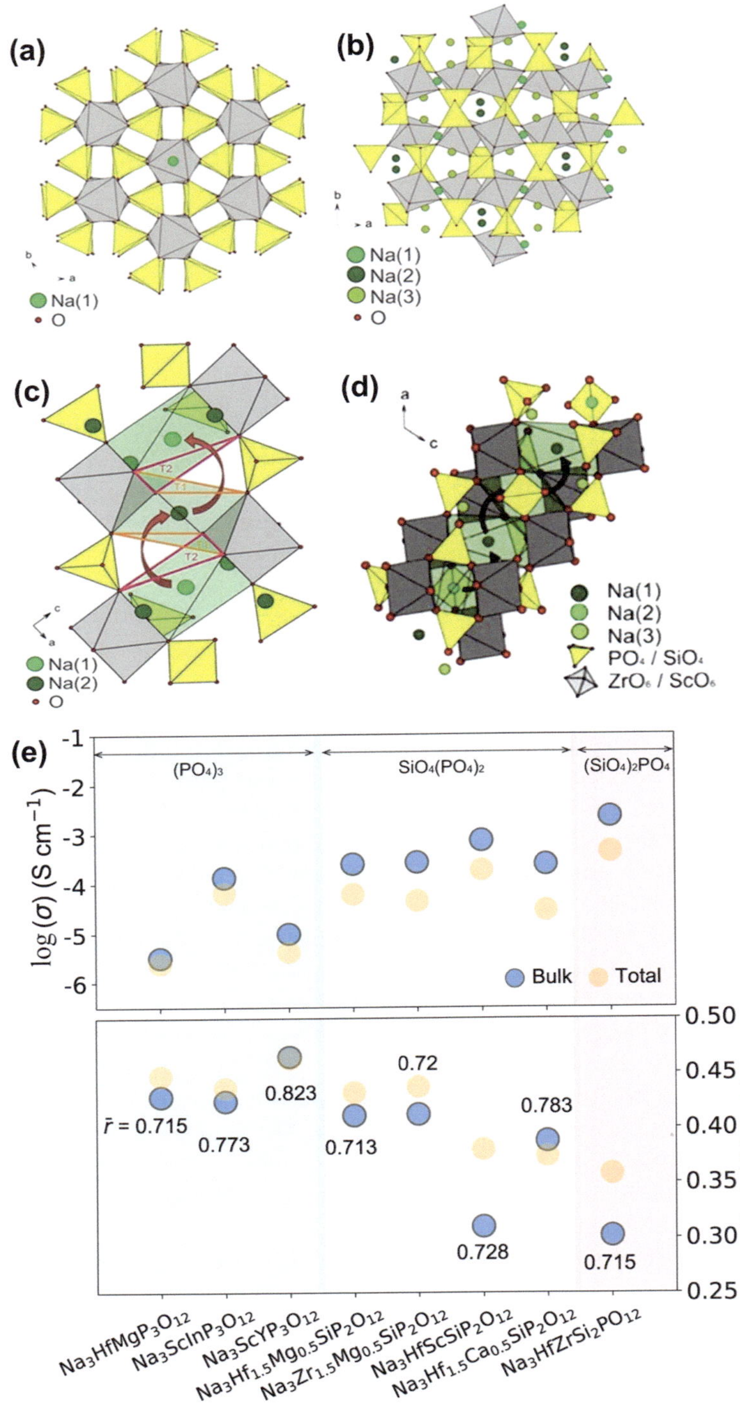

Fig. 2 NASICON-type structures **a** rhombohedral (R3c) and **b** monoclinic (C2/c) (the yellow tetrahedra and gray octahedra correspond to SiO_4/PO_4 and ZrO_6 units, respectively); Na-ion conduction channel in NASICON-type **c** rhombohedral and **d** monoclinic structures [38]. **e** Scatter plots of conductivity and activation energies at room temperature (~25 °C). On the x-axis, compounds are categorized by polyanion chemistry, and compounds in each group are sorted by increasing average metal radius [39]

only tetravalent ions M^{4+} (M = Ti, Ge, or Sn). To increase the ionic conductivity of $LiM_2(PO_4)_3$, a common approach is to partly replace tetravalent ions by trivalent ions $(M)^{3+}$ (M = Al, Ge, In, La, etc.). The most prominent NASICON-type FICs are $Li_{3-2x}(Al_{1-x}Ti_x)_2(PO_4)_3$ (LATP) [40] and $Li_{3-2x}(Al_{1-x}Ge_x)_2(PO_4)_3$ (LAGP) [41], which have high bulk ionic conductivities in the range of 10^{-4} ~10^{-3} Ω^{-1} cm^{-1} at ambient temperature. However, NASICON's weak interfacial compatibility causes poor interface contact between SSEs and inorganic cathodes, which raises interfacial impedance and eliminates effective ion transport channels. Moreover, cycling would cause a volume shift in the composite cathode's active materials, jeopardizing the integrity of the composite cathode and interfacial contact. As a result, the high interfacial resistance will lead to polarization and lower capacity, as well as poor cycle and rate performance [42].

2.4 Composite Polymer Electrolyte (CPEs)

The past several decades have seen a significant amount of research on polymers, starting with the discovery in 1973 [43] that polyethylene oxide (PEO) was suitable as an electrolyte for batteries when combined with alkali metal salts. As an electrolyte membrane, a wide range of polymers such as polymethyl methacrylate (PMMA), polyvinylidene fluoride (PVDF), and polyacrylonitrile (PAN) arose [44]. However, polymer-alkali metal salt membranes have several shortcomings that drive the search for improvement, such as low mechanical strength, poor film-forming ability, limited ionic conductivity, thermal instability, and flexibility [45]. Among the challenges, one of the most essential is the electrolyte's poor ionic conductivity, which must be addressed in order to make it acceptable for solid-state batteries. To overcome these issues, many have approached toward composite electrolytes made of organic polymers and inorganic ceramic filler. Ceramic fillers can be divided into two types: inert fillers (TiO_2, SiO_2, Al_2O_3) and active fillers (NASICON, Garnet, halide, etc.) as shown in Fig. 3 [46]. Both active and inert fillers can reduce the crystallinity of the polymer and allow Li^+ ions to migrate more freely. Active fillers, which are rapid ion conductors, can enhance ionic conductivity by promoting Li^+ diffusion across the permeation network created by continuous filler particles [47].

Inert Filler

The inclusion of inorganic fillers can lower the T_g while also reducing the tendency of the polymer to crystallize. In addition, these inorganic fillers possess surface groups that can interact with salts and polymer chains to promote faster ion migration on the inorganic surfaces via conductive pathways. These two factors together account for the majority of the reasons why the dispersion of inert fillers, such as TiO_2, SiO_2, Al_2O_3, ZrO_2, and others, improves the ionic conductivity of electrolytes [48]. At room temperature, the dispersion of inert filler Al_2O_3 has been reported to have an ionic conductivity of around $9 \times 10^{-5}\ \Omega^{-1}\ cm^{-1}$ [48]. However, the conductivity improvement is limited to some extent due to the quantity and size of fillers. Insulating fillers with a high content ($\geq 15\%$) operate as barriers to the movement of mobile ions, reducing overall ionic conductivity.

Active Filler

The dispersion of FIC fillers in CPEs has been attempted recently, since the conductivity increase in CPEs is limited by the number and size of fillers. Because of three factors: (1) movement of the polymer segments; (2) vacancy or interstitial hoping in the ceramic fillers; and (3) migration at the interface of polymers and fillers, Li^+/Na^+ ion FICs as active fillers may enhance the electrochemical performance of composite solid polymer electrolytes more successfully than inert fillers [49, 50]). In general, FICs include garnet-type ($Li_7La_3Zr_2O_{12}$, $Li_{6.4}La_3Zr_{1.4}Ta_{0.6}O_{12}$, etc.) NASICON-type ($Na_3Zr_2Si_2PO_{12}$, $Li_{1.3}Al_{0.3}Ti_{1.7}(PO_4)_3$, perovskite-type ($Li_{0.35}La_{0.55}TiO_3$), sulfide-type materials ($Li_{10}GeP_2S_{12}$, $Li_2S–P_2S_5$, Li_6PS_5X), and halide-type materials (Li_3YCl_6, Li_3ScCl_6, Li_3YBr_6) [26, 29, 51–54].

With the dispersion of FICs fillers, ionic conductivity in the ranges of ~10^{-4} to $10^{-3}\ \Omega^{-1}\ cm^{-1}$ at room temperature with an appreciable electrochemical stability window has been achieved with various combinations of polymer with alkali metal salts and ceramic fillers (Fig. 5a) [42]. Dispersion of active fillers in polymer not only enhanced ionic conductivity

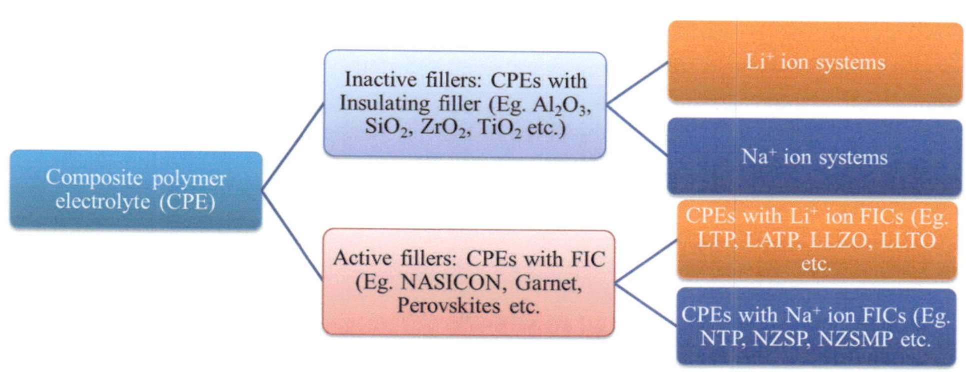

Fig. 3 Types of fillers used in composites polymer electrolytes (CPEs). FIC: Fast ionic conductor

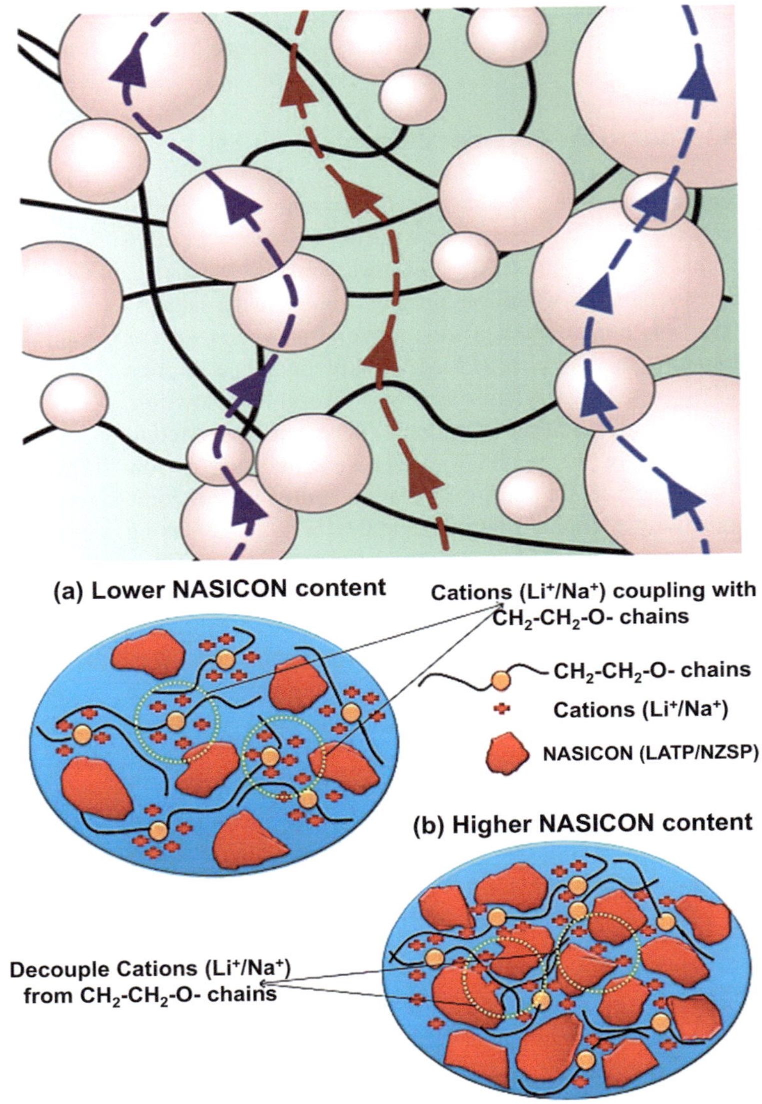

Fig. 4 Ionic transport mechanism in polymer-ceramic composite electrolytes [49]

but also electrochemical stability window and mechanical strength of the polymer composite electrolytes.

Despite the large electrochemical stability window of composite polymer electrolytes along with good interface-forming ability with electrodes, the room temperature ionic conductivity of the electrolyte, as illustrated in Fig. 5b, must be addressed.

3 Challenges in SSBs

3.1 Poor Mechanical Properties of SSEs

Lithium dendrites, crystalline structures formed under non-equilibrium conditions, exhibit a typical dendritic morphology, as depicted in Fig. 6 and reported by [55]. Research indicates that the growth of lithium dendrites is largely influenced by current density and operating temperature. As a result of a thinner SEI layer, excessive current density and mild working temperature might raise mass transport resistance of the lithium-ion and decrease the resistance of surface reaction, which can encourage the development of lithium dendrites [56, 57]. Li dendrite growth poses safety concerns, and efforts to mitigate this issue involve strategies such as using SSEs [58].

In 1976, Hong and Goodenough et al. introduced NASICON-based SSEs, specifically $Na_{1+x}Zr_2P_{3-x}Si_xO_{12}$ $(0 \leq x \leq 3)$, known for their high conductivity facilitated by open 3D channels, making them promising for sodium-ion SSBs (SSSBs) [35, 59]. While modeling by Newman and Monroe suggested potential Li dendrite evolution mitigation through SSEs if the shear modulus was approximately

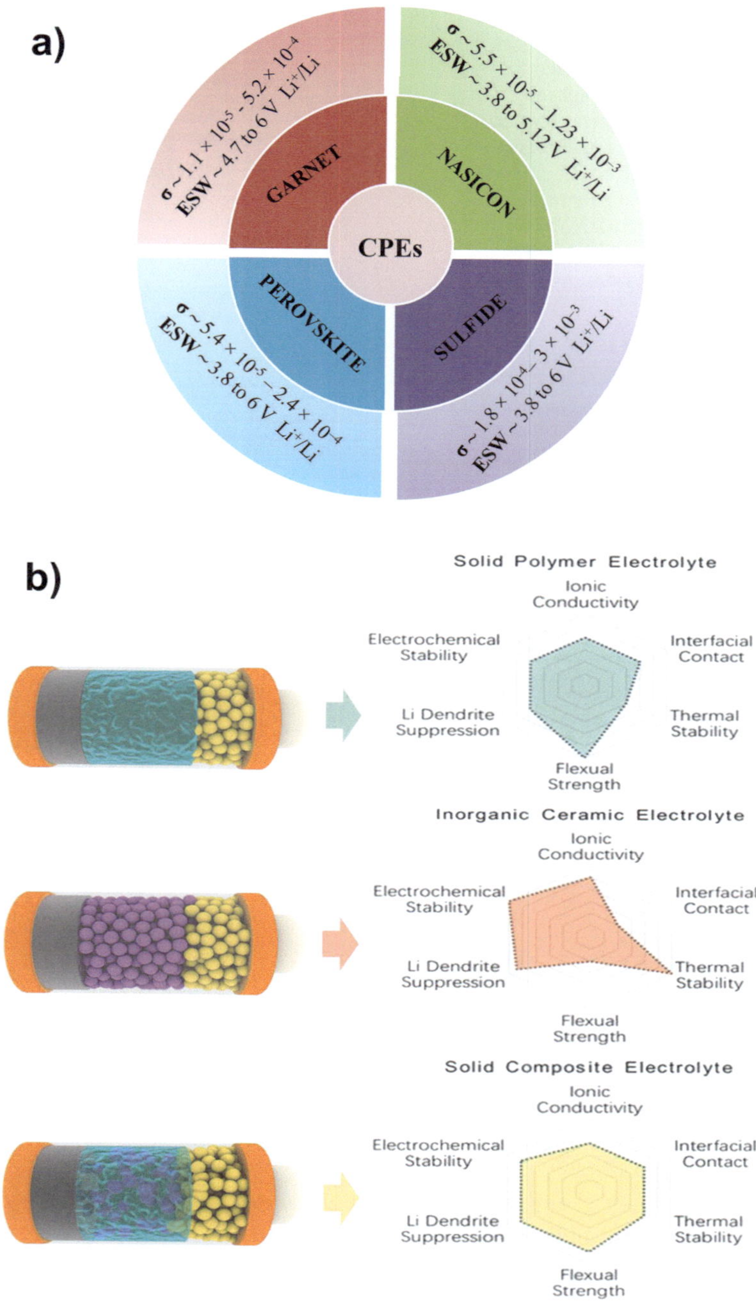

Fig. 5 **a** Ionic conductivity ranges of polymer-ceramic composites **b** The performance comparisons of solid polymer electrolytes, ceramic electrolytes, and composite polymer electrolyte [42]

double than that of lithium [60], NASICON-based SSEs, with a higher shear modulus (47.7 GPa) compared to Na metal (3.3 GPa), were expected to resist Na dendrites [61]. However, Na dendrites even instigated short circuits in NASICON-based SSSBs [62]. Various interfacial engineering methods, such as greater uniaxial compression [62], nanomaterial composite anodes [63], 3D sodium metal anodes, plated interlayers [64], in situ chemical developing interphase [65], and embedding a polymer SSE interlayer [64], have been explored to suppress Na dendrites.

Due to inadequate contact, dendrites can grow not merely at the electrolyte/anode interface, but also at the electrolyte pores and grain boundaries. Metal deposition at the anode occurs when the electrolyte grain boundary shows higher impedance than the grain [67]. Pores generated during the electrolyte calcination also induce metal deposition at the anode side [68]. Partially substituting Zr^{4+} in $Na_3Zr_2Si_2PO_{12}$ electrolyte with heterogeneous ions has been acknowledged as a successful method for raising relative density and ionic conductivity [69]. For

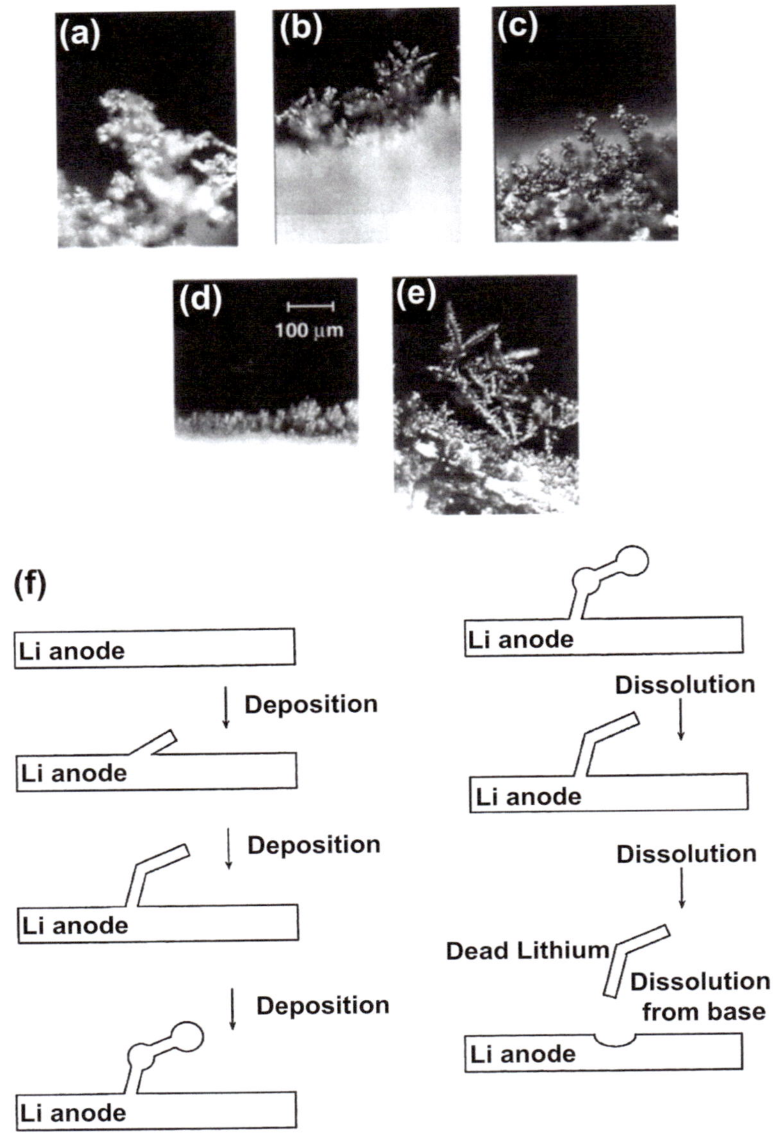

Fig. 6 After the electrochemical deposition of lithium (3 mA cm^{-1} × 1 h) from PMMA-based electrolytes, micrographs of titanium surfaces were obtained. The corresponding molar ratios for triethylene glycol dimethacrylate to polymethyl methacrylate are **a** 0%, **b** 1%, **c** 2%, **d** 4%, and **e** 8%, f. The dissolution and deposition model of lithium [66]

instance, Lu et al. enhanced Na^+ mobility by doping Ca^{2+}, resulting in an ionic conductivity of 1.6×10^{-3} S cm^{-1} and stable operation for 600 h at room temperature [64]. Yang et al. employed Mg^{2+} as a dopant ion, creating a positive charge deficiency to increase Na^+ concentration, achieving a conductivity of 1.3×10^{-3} S cm^{-1} [70]. By adding La^{3+} to create $Na_3La(PO_4)_2$, Zhang et al. were able to increase relative density and improve Na^+ concentration, which resulted in an ionic conductivity of ~3.4×10^{-3} S cm^{-1}. At a high current density of 1.1 $mAcm^{-2}$, sodium metal symmetrical cells showed good cycling performance, indicating the enhancement of NASICON critical current density (CCD) by ion replacement [49]. However, understanding the mechanisms behind Na dendrite formation and propagation in NASICON-based SSEs is still incomplete, emphasizing the need for comprehensive studies, especially focusing on optimizing the SEI for stable SSSB operation. Direct observations are critical for developing effective pathways for suppression, which is lacking in SSEs, necessitating comprehensive studies to advance their application in SSSBs.

Garnet-based solid-state lithium metal batteries, known for their improved energy density and safe operation, face challenges in practical applications despite the high shear modulus of Garnet SSEs (GSSEs). The precise mechanism governing Li dendrite growth is still unclear, and is attributed to factors such as imperfect contact, contaminants, grain boundaries, and high electronic conductivity within the GSSE. Changes in the Li-ion and electron distribution at the electrode/GSSE interface and within the GSSE bulk/grain boundaries are critical to prevent the growth of Li dendrites. Various approaches have been suggested, such as anode modification, composite electrolytes, bulk optimization, interfacial engineering, and electrochemical pre-treatments [71]. Recent studies emphasize the importance of lowering electronic conductivity for the accomplishment of ASSBs. Han et al. demonstrated that the high electronic conductivity of LLZO contributed to dendrite formation [72]. Similarly, Ping et al. studied the short-circuit behavior of a garnet electrolyte at 60 °C and inferred that enhancing the ionic conductivity and suppressing the electronic conductivity were important steps to prevent short circuits in solid-state batteries [73].

To comprehend the mechanism behind the growth of Li dendrites, various factors influencing their growth have been investigated. Figure 7a and b illustrate superficial and grain boundary Griffith cracks within GSSE polycrystals, serving as catalysts for Li dendrite growth. The electric field near these cracks exceeds that of other regions, intensifying the development of Li dendrites particularly in specific regions. The growth of lithium dendrites within Griffith cracks and grain boundaries initiates incessant and unrestricted progress, and as stress is released, the cracks and grain boundaries are subsequently filled. Once these grain boundaries and Griffith cracks permeate all over the GSSE, dendrite expansion initiates at these points, persisting until a cell short circuit occurs. Cheng et al. witnessed distinctive hexagonal mesh-like structure of dendrites, confirming this theory (Fig. 7c) [74].

In scenarios involving discrete Griffith cracks or grain boundaries, dendrites continue to grow alongside the expanding Griffith cracks or grain boundaries till the entire GSSE is penetrated by Li dendrites (Fig. 7d). Schmidt and Sakamoto demonstrated this process, monitoring the microstructural destruction of GSSE throughout charging and discharging by means of an in situ pulse-echo transduction technique [74]. With dendrite proliferation, the Griffith cracks and grain boundaries were filled(Fig. 7e). Voids within GSSEs also contribute to Li dendrite growth, leading to the expansion of cavities and dendrite infiltration within GSSEs (Fig. 7f). Aguesse et al. observed Li dendrite development in the clefts of GSSE, as depicted in Fig. 7g [75].

The growth of dendrites tends to favor areas with greater electric fields, for example, grain boundaries, Griffith cracks, and cavities within GSSEs [62]. Li dendrite growth is not limited by the Griffith fractures or voids, even in a single-crystal GSSE. Dendrites grow particularly from areas where Griffith fractures are present on the surface or cavities of single-crystal GSSE. These spots may extend and allow dendrites to infiltrate the single-crystal GSSE (Fig. 7h and 7i). Swamy et al. investigated this process, and the proposed

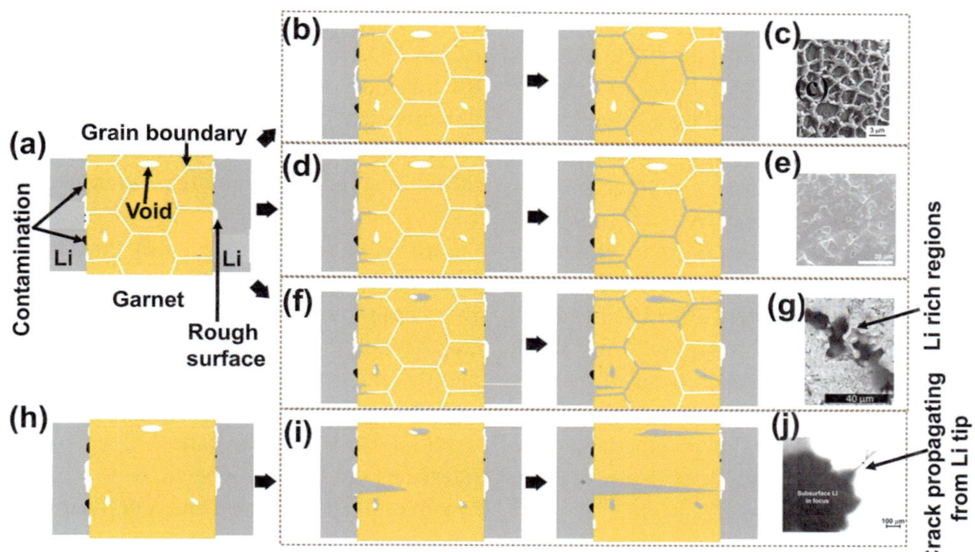

Fig. 7 The schematic in **a** outlines the GSSE polycrystals, while **b** details Li dendrite growth within the grain boundaries and Griffith cracks. An SEM image in **c** reveals the distinct mesh arrangement observed in GSSE after cycling, emphasizing the intricate nature of Li dendrite formation. Schematics in **d** depict Li dendrite growth concurrent with the expansion of grain boundaries and Griffith cracks in GSSE. Further insights are gained from the SEM image in **e**, displaying the cracked and glossy surface of GSSE post-cycling, highlighting the intersected spongy structure [74]. **f** illustrates Li dendrite growth originating from the voids within GSSE polycrystals, while **g** presents an SEM image of a fissured pellet, showcasing Li dendrite growth inside the pellet [75]. Schematics in **h** and **i** elaborate on GSSE single crystals and Li dendrite growth within them. The transverse view ex situ optical microscope image obtained from a post-short circuit in **j** provides a closer look at the Li dendrite growth in a GSSE single crystal [76]. This compilation offers valuable insights into the diverse scenarios influencing Li dendrite growth within garnet structures

mechanism for the propagation of cracks during Li dendrite growth is illustrated in Fig. 7j [76].

The increasing need for LIBs with extraordinary energy density highlights the importance of polymer-based solid-state electrolytes (PSSEs). These PSSEs offer a favorable combination of thinner and lighter structures, facilitating superior electrode/electrolyte interfacial contact. Gel polymer electrolytes (GPEs), polymer electrolyte composites (CPEs), and ceramic-filled composite gel polymer electrolytes (CGPEs) are the products of the introduction of various additives or fillers, such as liquid electrolytes (LEs), ionic liquids (ILs), and inorganic ceramic electrolytes (ICEs), to improve the electrochemical, thermal, and mechanical stability of PSSEs [46].

The unequal accumulation of Li-ion at the interface, which causes the production of dendrites and seriously jeopardizes the cycle performance and safety of solid-state LIBs (SSLIBs), is a major cause for concern in polymer-based SSEs. This problem is particularly noticeable in ultra-thin polymer-based SSEs with high energy density SSLIBs. Although the addition of ceramic fillers to CPEs and CGPEs results in improved mechanical strength, ionic conductivity may be compromised by the severe agglomeration of ceramic fillers in high-content "polymer-in-ceramic" CPE structures. To address this, a bilayer electrolyte design with a mechanically robust layer containing 40 wt.% LLZTO placed toward the lithium metal anode has been proposed. This design demonstrated stable cycling for 400 h, outperforming a single-layer electrolyte with 10 wt.% LLZTO. Furthermore, the effectiveness of ceramic fillers to suppress lithium dendrites can be further improved by changing their shape, for example, by adding nanowires or creating a ceramic network inside CPEs [77]. Novel approaches, like in situ polymerization of ϑ-caprolactone (ϑ-CL) by polymerization due to ring-opening, have been used to efficiently suppress lithium dendrites and create interphase layers with increased mechanical strength. Li ǁ PC GPE/PCL-LiTFSI ǁ Li cells, for example, cycled at 0.05 mAh cm^{-2} for 900 h at 60 °C showed stable performance accompanied by only a minor increase in overpotential. In contrast, following a 200 h cycle, a cell containing an ex situ polymerized electrolyte suffered a short circuit after 200 h of cyclic plating-deplating process [77].

Sulfide solid electrolytes, particularly lithium thiophosphate solid electrolytes (TSEs), have gained interest for ASSBs due to their formability and high ionic conductivity at room temperature. Despite some TSEs exhibiting ionic conductivities comparable to or surpassing liquid electrolytes, challenges arise from inadequate mechanical and electrochemical stabilities, resulting in increased interfacial resistance. Furthermore, TSEs are unstable against lithium metal, leading to interphases that elevate localized current density at the Li/TSEs interface and encourage the lithium dendrite formation. Incorporating a conducting layer for li-ion, which must be insulating electronically at the Li/TSEs interface, enhances effective contact area and reduces electronic conductivity of the SEI and electrolyte layers [53]. Utilizing LiF, a crucial SEI layer component, proves effective in suppressing lithium dendrite growth. Ionic liquids like LiTFSI/Pyr$_{13}$TFSI have been employed for in situ LiF layer formation to protect the anode of Li metal [54]. Various inorganic SEI layers, including Li_xSiS_y [78], LiI [66], and LiH_2PO_4 [79], have been proposed, but their incremental improvements in critical current density have been insufficient. Utilizing molecular layer deposition technology, an organic–inorganic composite SEI with enhanced mechanical characteristics and the ability to adapt to variations in stress and strain has been created on lithium metal surfaces, improving cycling performance. Long cycle life with homogeneous lithium plating/stripping is ensured by optimizing TSE composition, such as via halide-doping with elements like F/I [80].

3.2 Anode Electrolyte Interface

One of the major challenges in developing solid-state batteries is to ensure a stable and efficient interface between the anode and the electrolyte. The anode–electrolyte interface is prone to chemical and electrochemical reactions that can degrade the performance and safety of the battery. These reactions can result in the formation of interphase layers that can affect the transport of ions and electrons, as well as the growth of dendrites that can cause short circuits. Therefore, it is important to understand the mechanisms and factors that influence the interface stability and to design strategies to mitigate these issues.

In a recent review [81] identify three main types of interactions that occur at the anode–electrolyte interface: (a) contact, (b) charge transfer, and (c) Li dendrite formation (Fig. 8a). Contact refers to the physical contact between the anode and the electrolyte, which can be influenced by the surface morphology, roughness, porosity, and wettability of both materials. In solid-state batteries, typically, rigid solid–solid interfaces are formed wherein a good contact can ensure a low interfacial resistance and a uniform current distribution, while a poor contact can lead to high polarization and local hot spots [82]. Charge transfer refers to the electrochemical processes that take place at the interface, such as Li-ion insertion/extraction and electron transfer. These processes are heavily influenced by side reactions, reaction kinetics, interfacial evolution, and the dynamic interfacial resistance. They can determine the kinetics and thermodynamics of the battery operation, as well as the stability and compatibility of the materials [83, 84]. Another consequence of spontaneous chemical reactions at the interface is the formation

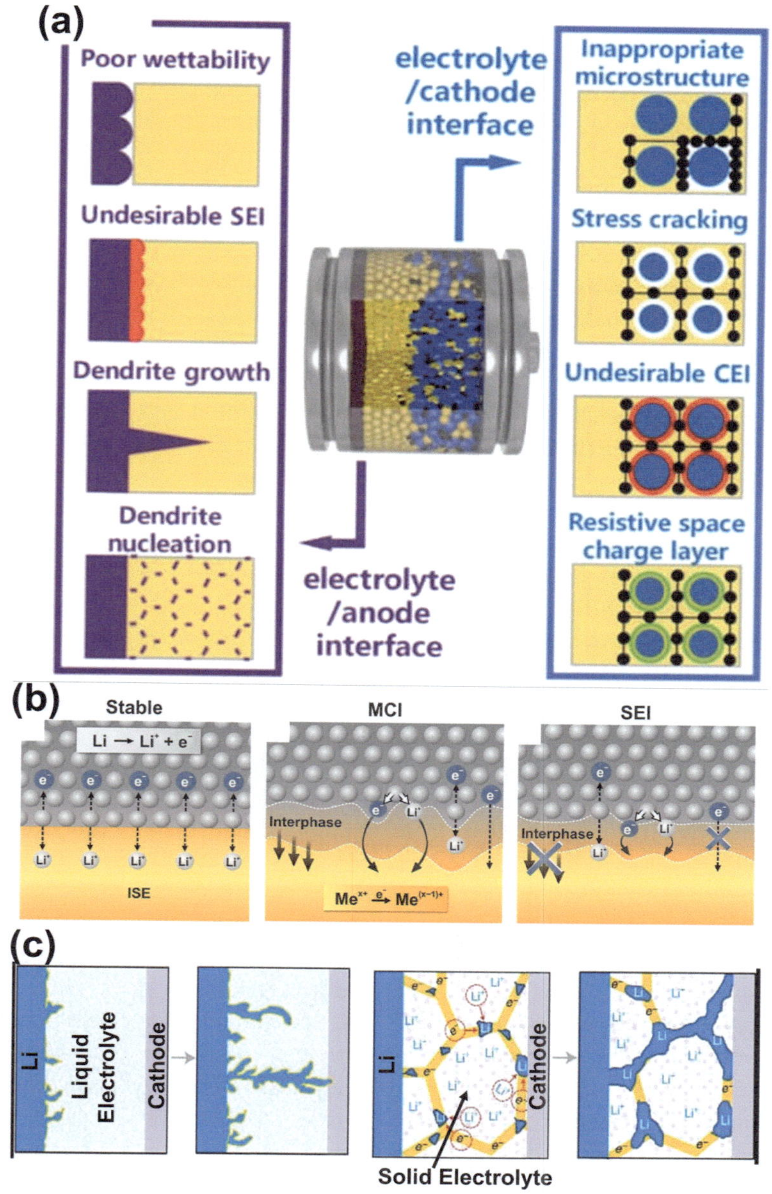

Fig. 8 Schematic depicting **a** prominent issues occurring at the electrode–electrolyte interfaces in state-of-the-art all-solid-state batteries [81] Reproduced with permission from Springer Nature, **b** the different types of interfaces/ interphases forming at the interface of Li metal with inorganic solid-state electrolytes (ISE)—thermodynamically stable interface, mixed-ionic/electronic conducting interphase (MCI), and a kinetically stabilized solid-electrolyte interphase (SEI) that is largely similar to the SEI formed in liquid-electrolyte-based systems (Reprinted with permission from [85]. Copyright 2020 American Chemical Society), and **c** Li filament formation and evolution across the interface between Li metal and inorganic electrolytes, distinguished from the formation of Li dendrites in liquid-electrolyte containing cells [88] Reproduced with permission from Springer Nature

of a solid-electrolyte interphase, or SEI (Fig. 8b), which could be thermodynamically stable, or mixed-ionic/electronic conducting, or metastable, i.e., kinetically stabilized [83, 85, 86]. While a thermodynamically stable or kinetically stabilized interphase is beneficial, forming a passivating layer that is Li-ion conducting but electronically insulating, potentially extending the window of the solid electrolyte's working voltage. Finally, Li dendrite formation refers to the growth of Li metal filaments from the anode into the electrolyte, which can be driven by various factors such as current density, temperature, stress, and concentration gradient. Li dendrites can penetrate through the solid electrolyte and reach the cathode, causing a short circuit or even a fire hazard, as depicted in Fig. 8c [87].

The solid-electrolyte interphase (SEI) formed at the anode-electrolyte interface is critical to its stability, influencing the electrochemical performance, cyclic charge–discharge processes, and the safety of the overall device. Unlike in liquid-electrolyte systems, the SEIs in SSBs are further complicated by the involvement of solid–solid interfaces, where mechanical mismatch, chemical instability, and ionic transport barriers must also be accounted for. Overcoming issues such as dendrite penetration in lithium metal anodes and resistive interphases with graphite necessitates the formation of a robust SEI. Sulfide-based electrolytes, such as Li_6PS_5Cl (LPSC) and $Li_{10}GeP_2S_{12}$ (LGPS), are characterized by their high ionic conductivity but suffer from chemical instability when in contact with lithium metal anodes. Studies indicate that such sulfides typically form an SEI consisting of Li_2S, Li_3P, and polysulfides, which are ion-conductive but prone to decomposition under cycling conditions.[89, 90] Whereas oxide-based systems, particularly garnet-type solid electrolytes (e.g., LLZO), offer excellent chemical stability against lithium metal, forming a highly stable SEI, but face challenges such as poor wettability, high interfacial impedance, and dendrite growth [81]. Meanwhile, polymer-based solid electrolytes (e.g., PEO) are attractive for their flexibility and processing advantages but exhibit limited ionic conductivity and poor stability with lithium metal. Composite structures combining polymers with inorganic electrolytes have been shown to improve interfacial compatibility and SEI uniformity. Recent findings suggest that a polymer–solid hybrid interphase can form a mixed organic–inorganic SEI to enhance lithium-ion transport and mitigate side reactions [91, 92].

These studies have driven intensive research aimed at characterizing SEIs, understanding their dynamic evolution, and engineering interfaces to optimize their properties (Fig. 9a) [91, 93–95]. Furthermore, to enhance the anode–electrolyte interface, [81] suggest two main approaches: compositional modifications and structural designs (Fig. 9b). Compositional modifications involve changing the chemical composition or adding additives to either the anode or the electrolyte to improve their compatibility, conductivity, stability, or reactivity. For example, adding Li salts, ceramic fillers, or polymers to solid electrolytes can increase their Li-ion conductivity, mechanical strength, or flexibility [82, 95]. Adding carbon, metal, or alloy coatings to anodes can improve their electronic conductivity, surface protection, or metal plating/stripping behavior [81, 84]. Structural designs involve changing the shape, size, arrangement, or architecture of either the anode or the electrolyte to optimize their contact area, interfacial resistance, or stress distribution. For example, using thin films, nanowires, porous structures, or 3D networks can enhance the anode–electrolyte interface by increasing their surface area, reducing their thickness or creating more pathways for ion/electron transport [87]. In this manner, interfacial layers, or interlayers, play a critical role in stabilizing the solid-electrolyte interphase (SEI), improving electrochemical stability at the interface, and enhancing solid–solid contact between the anode and the solid-state electrolyte (SSE). They achieve this by suppressing undesirable interfacial reactions, reducing interfacial resistance, improving ionic and electronic conductivity, and mitigating mechanical and thermal stresses.

Various types of interlayer materials have been proposed, including metallic coatings, lithiophilic layers, artificial SEI layers, and polymer-based buffer interlayers. In sulfide-based systems, interlayers suppress undesirable reduction reactions and form protective SEIs, such as those rich in LiF, as demonstrated by Gao et al. [96], achieving stable lithium cycling over 3000 h. Polymer-sulfide hybrid interlayers, such as PEO-based interphases, improve ionic transport and mitigate interfacial resistance by forming mixed organic–inorganic SEI layers [91]. In garnet systems, interlayers like Sr_3N_2-modified lithium metal [97] or ZnO coatings [98] improve wetting and reduce interfacial area-specific resistance to as low as 20 $\Omega.cm^2$. Lithiophilic metallic coatings, such as Ag or Li_3N interlayers, further stabilize the garnet–lithium interface, promoting homogeneous lithium flux and reducing dendritic growth [14, 94]. For polymer systems, inorganic-rich SEIs catalyzed by interfacial layers significantly enhance cycle life and prevent dendrite formation [92]. Across all systems, interlayers reduce interfacial impedance, enable controlled SEI growth, and mitigate mechanical or chemical failures, ensuring stable cycling in prototype and experimental SSBs.

However, these approaches may not be sufficient to solve all the interfacial issues in solid-state batteries. Some of the challenges that still need to be addressed include: (1) how to achieve a good contact between different types of solid electrolytes (e.g., sulfide, oxide, or polymer) and various anode materials (e.g., Li metal, graphite, or alloy); (2) how to prevent or minimize the side reactions between the anode and the electrolyte that can consume active materials or form undesirable products; (3) how to suppress or eliminate Li dendrite formation under different operating conditions (e.g., high current density or low temperature); and (4) how to characterize and monitor the interfacial phenomena in situ or operando with high spatial and temporal resolution [99].

In conclusion, the anode–electrolyte interface is a critical component in solid-state batteries that requires careful attention and engineering. By applying suitable compositional modifications and structural designs, it is possible to improve some aspects of the interface stability and efficiency. However, more fundamental understanding and innovative solutions are needed to overcome all the interfacial challenges and thus achieve better performance and safety of solid-state batteries.

3.3 Ambient Air Degradation of Electrolytes

While oxide SSEs are typically stable in ambient air, the susceptibility of garnet-type SSEs to CO_2 and moisture poses a challenge. This vulnerability results in deposition of Li_2CO_3 layer, resulting in the subsequent degradation of ionic conductivity. Comprehending the air stability of garnet SSEs is vital, and it is imperative to find effective solutions to prevent contamination [100]. In a study by Nyman et al., garnet-type powder samples were dispersed in water, leading to an observed rise in the water's pH value [101]. This observation

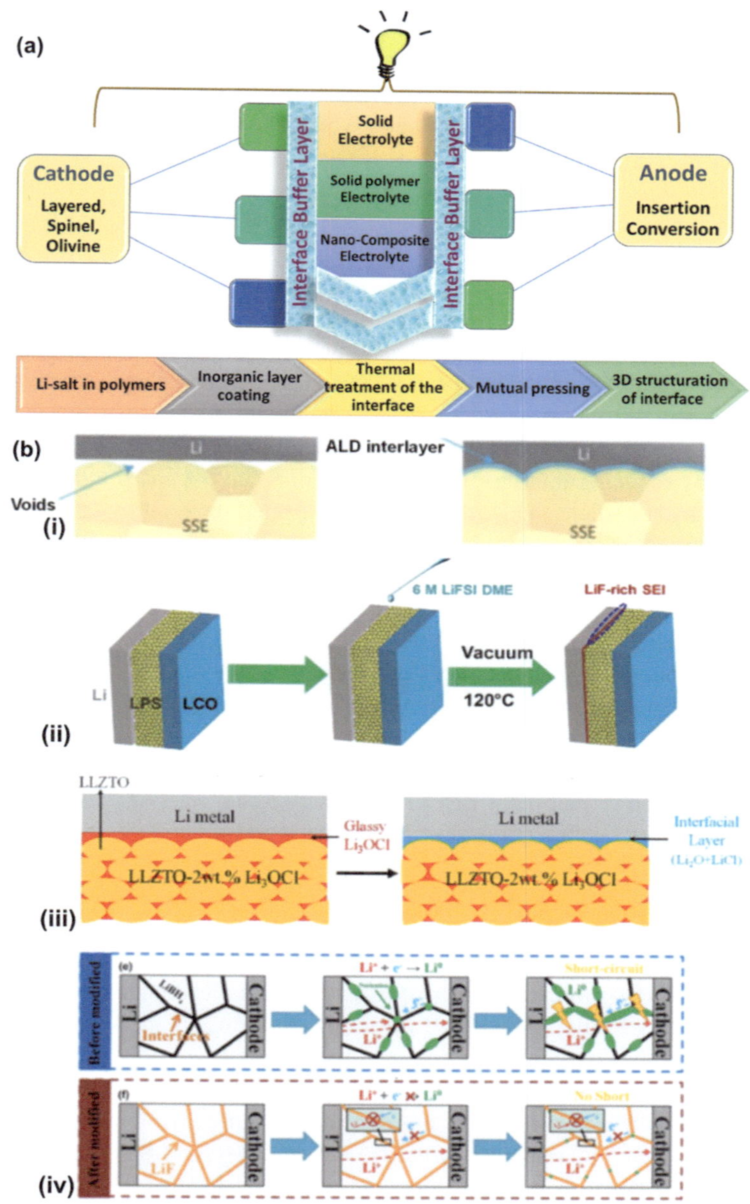

Fig. 9 (**a**) Approaches considered for minimizing interfacial resistance in all-solid-state cell configurations (Reprinted with permission from (Jena et al. 2018). Copyright 2018 American Chemical Society); **b** specific examples of such approaches, including (i) improved Li wetting at an ALD-modified surface, (ii) stabilization of SEI using a pre-formed LiF layer, (iii) an interlayer formed in situ, and (iv) doping/modification of electrolyte for suppressing Li filament formation [81] Reproduced with permission from Springer Nature

suggests that H^+ from water replaces Li^+ in the garnet structure. Galven et al. showcased the spontaneous exchange of Li^+/H^+ on the garnet surface, occurring at low temperatures (<150 °C) and under ambient air conditions. Additionally, their findings revealed a direct correlation between Li^+/H^+ exchange and lithium stoichiometry, as confirmed by powder neutron diffraction [102].

Cheng et al. examined the electrolyte subsequent to the polishing process and the 2-month storage of garnet SSEs in ambient air. Laser-induced breakdown spectroscopy unveiled a substantial Li/Zr layer on the surface of LLZO [103]. This observation was corroborated by synchrotron X-ray diffraction (XRD), X-ray photoelectron spectroscopy (XPS), and X-ray absorption spectrum (XAS) all validating the deposition of Li_2CO_3 on the LLZO.

The generation of Li_2CO_3 can occur through three distinct pathways. While some studies propose a one-stage reaction in which garnet reacts directly with CO_2 to produce Li_2CO_3, a more widely accepted two-step mechanism is suggested by Xia et al. In this mechanism, the garnet surface initially adsorbs water vapor, leading to Li^+/H^+ exchange and the formation of a layer of $LiOH{\bullet}H_2O$ [100]. Subsequently, this layer reacts with CO_2 to generate a Li_2CO_3 layer on the garnet surface. Density functional theory (DFT) calculations support this mechanism, despite thermodynamic stability with carbon-containing species, leading to slower kinetics.

Another pathway involves formation of LiOH via hydration without the exchange of Li + /H + , as shown in the equation $Li_7La_3Zr_2O_{12} + xH_2O \rightarrow Li_{7-2x}La_3Zr_2O_{12-x} + 2xLiOH$ [104]. Li_2CO_3 may degrade at higher voltage than 3.2 V by reaction $2Li_2CO_3 + C \rightarrow 4Li^+ + 3CO_2 + 4e^-$, disrupting the cathode/garnet interface and deteriorating the cathode-side interface. Therefore, protecting garnet-type SSEs from humid air is necessary.

In response to surface contaminants on garnet SSEs, Goodenough et al. proposed an effective approach to enhance stability in humid air [105]. This involved the addition of 2 wt% LiF to the garnet-type SSEs. Furthermore, they employed a straightforward method to eliminate Li_2CO_3 by subjecting the garnet to a reaction with carbon at 700 °C, thereby reducing interface resistance. To mitigate the sharp interface resistance caused by a contaminated layer, the researchers adopted a 3D structure with porous-dense-porous layers. This structural arrangement proves to be an efficient solution to address interfacial challenges. Additionally, the polishing of garnet SSEs in a glovebox filled with Argon emerged as a practical method to further manage surface contaminants and enhance the overall performance of the electrolytes. Cheng et al. recently discovered that smaller grains of Al-doped LLZO exhibit greater stability and reduced formation of Li_2CO_3 compared to larger grains [103]. This phenomenon is ascribed to the segregation of Al at grain boundaries in larger LLZO grains, leading to surfaces depleted of aluminum, whereas small LLZO grains retain Al, resulting in an aluminum-rich surface. Consequently, smaller Al-doped LLZO grains exhibit lower Li_2CO_3 formation on surfaces following exposure to ambient air, leading to enhanced stability and reduced interfacial resistance. Effective strategies for enhancing the stability of garnet-type SSE involve techniques such as microstructure modification, the introduction of sintering aids, and surface polishing.

The potential stability window of garnet-type SSEs has also raised concerns, alongside issues related to air stability. Han et al. highlighted discrepancies in measuring the thermal electrochemical stability using Li/electrolyte/inert metal electrodes, employing first-principles methods for calculation [51]. Addressing this issue, a Li/electrolyte/carbon-electrolyte arrangement was suggested to determine the inherent stability potential window effectively. Goodenough et al. [59] noted the influence of the Li_2CO_3 layer on electrochemical window measurements, asserting that garnet SSEs without this layer remained stable up to 4.4 V [105]. Hence, future research on garnet-type SSEs should accurately determine the electrochemical window, acknowledging the impact of factors like interface resistance and the presence of Li_2CO_3.

The vulnerability of sulfide solid-state electrolytes (SSEs) to air instability is the main obstacle to their commercialization. The capacity of a substance to retain its structure, chemical structure, as well as other characteristics without reacting with airborne constituents is referred to as air stability. When it comes to sulfide SSEs, the hydrolysis reaction with H_2O in humid air [106] causes air instability rather than reactions with N_2 [107] and CO_2 [108]. This hydrolysis reaction produces toxic H_2S gas, leading to structural collapse, a rapid reduction in ionic conductivity, and severely affects the safety [106]. As a result, sulfide SSE synthesis necessitates regulated, low-moisture environments, including glove boxes filled with Argon or extremely dry rooms. This limitation raises the cost of manufacturing materials and processing batteries in addition to limiting the performance of sulfide SSLIBs. Sulfide-based SSLIBs must effectively increase their production scale and resolve material compatibility concerns in order to be competitive [109, 110]. It is important to emphasize the need to improve air stability because maintaining a low humidity environment comes with an accompanying cost rise. The main steps in producing humidity-resistant sulfide SSEs are compositional tuning and structural control, which include surface shape and local structure.

An important factor to take into account is chemical stability, which indicates a material's reactivity throughout storage. The idea of electrochemical stability, which is relevant to battery functioning, is measured using the term "electrochemical window." Increasing the cut-off voltage for charge/discharge or improving the active material's specific capacity are two methods to raise the energy density of lithium batteries. Higher cut-off voltages in ASSLBs require an electrochemical window for the SSE that is sufficiently broad. Sulfide SSEs' electrochemical window has been thoroughly investigated, requiring accurate identification and resolution of discrepancies between the material intrinsic electrochemical window and its functioning in complete cells. Precise measurement of intrinsic properties is essential to determine appropriate scenarios for SSE materials. The dominating effects of anions on reduction and oxidation potential (usually fluoride > oxide > sulfide > nitride), the impact of cations on reduction potential, and the relationship between concentration of Li and oxidation/reduction potentials are some of the factors that restrict the electrochemical window of SSEs. Theoretical simulations reveal that sulfide SSEs often exhibit a restricted electrochemical window, which is limited by the reduction of cations and oxidation of S^{2-} [111]. Experimental findings, however, reveal sulfides with a larger electrochemical stability window than predicted, adding complexity to the understanding of sulfide SSEs from material creation to full cell application [112]. The complexity of different sulfide SSEs' behavior in real-world applications is further compounded by numerous reports showing that they show superior cycle stability than anticipated at operation voltages far higher than their theoretical breakdown voltage [32].

4 Conclusion

In conclusion, ASSBs represent a promising frontier in energy storage technology, offering enhanced safety, prolonged cycle stability, and potential advancements in energy density. The choice of electrolyte materials plays a pivotal role in determining the overall performance of ASSBs. Garnet-type electrolytes, such as LLZO, have garnered considerable attention for their high ionic conductivity and stability, making them suitable candidates for ASSBs. However, challenges such as interfacial issues and fabrication complexities need further exploration to harness their full potential. NASICON-type electrolytes, characterized by materials like $Li_{1.3}Al_{0.3}Ti_{1.7}(PO_4)_3$, exhibit favorable electrochemical properties. However, when used with electrodes, NASICONs exhibit a high interfacial resistance and are mechanically fragile. Sulfide electrolytes have high ionic conductivity and offer good interface with electrodes owing to their soft nature, but face hurdles like air instability and narrow electrochemical window. Polymer composite electrolytes, combining organic polymers with inorganic ceramics or glass ceramics, offer flexibility and scalable solution processability while having compatible interface with electrodes, but their room temperature ionic conductivity is low and needs to be improved for commercialization.

Though research on SSBs has advanced quickly thus far, there are still many unanswered questions. To address them, it is suggested that future efforts focus on the following areas in order to get high-energy SSBs to the commercial level. (1) The results obtained at the present lab size ought to be transferred into the industrial-scale realization. For instance, to fulfill the requirements for pack weight and lower the battery's realistic resistance, the thickness of the SE must be optimized and modified for use in the battery. It would necessitate extensive study into creating the procedures for fabricating thin SE films, thin-film deposition techniques, or solution-processed SE infiltration technologies. (2) More comprehensive characterization instruments that can accurately monitor the interaction at interface during lithium plating/deplating close to SE are needed. Depending on the different types of electrolytes and operating conditions, the hidden dynamic lithium development can be clearly revealed by in-operando monitoring of how lithium nucleates and expands around imperfections at the interface or grain boundaries in the bulk of the SE. Furthermore, the ability to visualize at the atomistic level in situ would facilitate the unraveling of the migration mechanism of lithium in the anode and interphase characteristic on different SEs. This would aid in the understanding of how different tactics might improve the performance of lithium metals. (3) It is necessary to keep looking for new ways to generate SEs with high ionic conductivity. To stabilize the interface with Li anode without producing the mixed conducting interphase, SE must also possess ductility and softness that can buffer structural changes of anode during Li deposition/stripping.

References

1. Nishi Y (2001) Lithium ion secondary batteries; Past 10 years and the future. J Power Sources 100:101–106. https://doi.org/10.1016/S0378-7753(01)00887-4
2. Tarascon J (2001) Issues and challenges facing rechargeable lithium batteries. Nature 414:359–367. https://doi.org/10.1038/35104644
3. Owens BB, Skarstad PM (1992) Ambient temperature solid state batteries. Solid State Ion 53–56:665–672. https://doi.org/10.1016/0167-2738(92)90444-T
4. Forsyth M, Porcarelli L, Wang X et al (2019) Innovative electrolytes based on ionic liquids and polymers for next-generation solid-state batteries. Acc Chem Res 52:686–694. https://doi.org/10.1021/acs.accounts.8b00566
5. Manthiram A, Yu X, Wang S (2017) Lithium battery chemistries enabled by solid-state electrolytes. Nat Rev Mater 2:16103. https://doi.org/10.1038/natrevmats.2016.103
6. Xu S, Sun Z, Sun C et al (2020) Homogeneous and fast Ion conduction of PEO-Based solid-state electrolyte at low temperature. Adv Funct Mater 30:1–9. https://doi.org/10.1002/adfm.202007172
7. Singh B, Ghosh S, Aich S, Roy B (2017) Low temperature solid oxide electrolytes (LT-SOE): a review. J Power Sources 339:103–135. https://doi.org/10.1016/j.jpowsour.2016.11.019
8. Zhu L, Zhu P, Fang Q et al (2018) A novel solid PEO/LLTO-nanowires polymer composite electrolyte for solid-state lithium-ion battery. Electrochim Acta 292:718–726. https://doi.org/10.1016/j.electacta.2018.10.005
9. Li L, Duan H, Li J et al (2021) Toward high performance all-solid-state lithium batteries with high-voltage cathode materials: design strategies for solid electrolytes, cathode interfaces, and composite electrodes. Adv Energy Mater 11:1–32. https://doi.org/10.1002/aenm.202003154
10. Yao X, Huang B, Yin J, et al (2015) All-solid-state lithium batteries with inorganic solid electrolytes: review of fundamental science. Chin Phys B 25. https://doi.org/10.1088/1674-1056/25/1/018802
11. Chen CH, Xie S, Sperling E et al (2004) Stable lithium-ion conducting perovskite lithium-strontium-tantalum- zirconium-oxide system. Solid State Ion 167:263–272. https://doi.org/10.1016/j.ssi.2004.01.008
12. Zhang Y, Xie MX, Zhang W et al (2020) Synthesis and purification of SiS2 and Li2S for Li9.54Si1.74P1.44S11.7Cl0.3 solid electrolyte in Lithium-ion batteries. Mater Lett 266:9–12. https://doi.org/10.1016/j.matlet.2020.127508
13. Dzwonkowski P, Eddrief M, Julien C, Balkanski M (1991) Electrical a.c. conductivity of B2O3xLi2O glass thin films and analysis using the electric modulus formalism. Mater Sci Eng, B 8:193–200. https://doi.org/10.1016/0921-5107(91)90038-W
14. Cai M, Lu Y, Su J et al (2019) In Situ Lithiophilic Layer from H+/Li+ exchange on garnet surface for the stable lithium-solid electrolyte interface. ACS Appl Mater Interfaces 11:35030–35038. https://doi.org/10.1021/ACSAMI.9B13190/ASSET/IMAGES/LARGE/AM9B13190_0007.JPEG

15. Kennedy JH, Yang Y (1987) Glass-forming region and structure in SiS2Li2SLiX (X = Br, I). J Solid State Chem 69:252–257. https://doi.org/10.1016/0022-4596(87)90081-8
16. Prasada Rao R, Seshasayee M (2006) Molecular dynamics simulation of ternary glasses Li2S-P2S5-LiI. J Non Cryst Solids 352:3310–3314. https://doi.org/10.1016/j.jnoncrysol.2006.04.018
17. Takada K, Aotani N, Iwamoto K, Kondo S (1995) Electrochemical behavior of LixMO2 (M = Co, Ni) in all solid state cells using a glass electrolyte. Solid State Ion 79:284–287. https://doi.org/10.1016/0167-2738(95)00075-H
18. Wang D, Zhong G, Pang WK et al (2015) Toward understanding the lithium transport mechanism in garnet-type solid electrolytes: Li+ Ion exchanges and their mobility at octahedral/tetrahedral sites. Chem Mater 27:6650–6659. https://doi.org/10.1021/acs.chemmater.5b02429
19. Zhang Q, Cao D, Ma Y et al (2019) Sulfide-based solid-state electrolytes: synthesis, stability, and potential for all-solid-state batteries. Adv Mater 31:1–42. https://doi.org/10.1002/adma.201901131
20. Kamaya N, Homma K, Yamakawa Y et al (2011) A lithium superionic conductor. Nat Mater 10:682–686. https://doi.org/10.1038/nmat3066
21. Kato Y, Hori S, Saito T et al (2016) High-power all-solid-state batteries using sulfide superionic conductors. Nat Energy 1:1–7. https://doi.org/10.1038/nenergy.2016.30
22. Subramanian K, Alexander GV, Karthik K et al (2021) A brief review of recent advances in garnet structured solid electrolyte based lithium metal batteries. J Energy Storage 33:102157. https://doi.org/10.1016/j.est.2020.102157
23. Murugan R, Thangadurai V, Weppner W (2007) Fast lithium ion conduction in garnet-type Li7La 3Zr2O12. Angew Chem-Int Ed 46:7778–7781. https://doi.org/10.1002/anie.200701144
24. Xu M, Park MS, Lee JM et al (2012) Mechanisms of Li + transport in garnet-type cubic Li 3+xLa 3M 2O 12 (M = Te, Nb, Zr). Phys Rev B Condens Matter Mater Phys 85:1–5. https://doi.org/10.1103/PhysRevB.85.052301
25. Stallworth PE, Fontanella JJ, Wintersgill MC et al (1999) NMR, DSC and high pressure electrical conductivity studies of liquid and hybrid electrolytes. J Power Sources 81–82:739–747. https://doi.org/10.1016/S0378-7753(99)00144-5
26. Samson AJ, Hofstetter K, Bag S, Thangadurai V (2019) A bird's-eye view of Li-stuffed garnet-type Li7La3Zr2O12 ceramic electrolytes for advanced all-solid-state Li batteries. Energy Environ Sci 12:2957–2975. https://doi.org/10.1039/c9ee01548e
27. Cussen EJ (2010) Structure and ionic conductivity in lithium garnets. J Mater Chem 20:5167–5173. https://doi.org/10.1039/b925553b
28. O'Callaghan MP, Cussen EJ (2007) Lithium dimer formation in the Li-conducting garnets Li 5+xBaxLa3-xTa2O12 ($0 < x \leq 1.6$). Chem Commun 12:2048–2050. https://doi.org/10.1039/b700369b
29. Han J, Zhu J, Li Y et al (2012) Experimental visualization of lithium conduction pathways in garnet-type Li7La3Zr2O12. Chem Commun 48:9840–9842. https://doi.org/10.1039/c2cc35089k
30. Wagner R, Redhammer GJ, Rettenwander D et al (2016) Crystal structure of garnet-related Li-Ion conductor Li7-3xGaxLa3Zr2O12: fast Li-Ion conduction caused by a different cubic modification? Chem Mater 28:1861–1871. https://doi.org/10.1021/acs.chemmater.6b00038
31. van den Broek J, Afyon S, Rupp JLM (2016) Interface-engineered all-solid-state Li-Ion batteries based on garnet-type fast Li+ conductors. Adv Energy Mater 6:1–11. https://doi.org/10.1002/aenm.201600736
32. Chen R, Li Q, Yu X et al (2020) Approaching practically accessible solid-state batteries: stability issues related to solid electrolytes and interfaces. Chem Rev 120:6820–6877. https://doi.org/10.1021/acs.chemrev.9b00268
33. Huang X, Liu C, Lu Y et al (2018) A Li-Garnet composite ceramic electrolyte and its solid-state Li-S battery. J Power Sources 382:190–197. https://doi.org/10.1016/j.jpowsour.2017.11.074
34. Yi M, Liu T, Li J et al (2019) High Li-ion conductivity of Al-free Li 7–3x Ga x La 3 Zr 2 O 12 solid electrolyte prepared by liquid-phase sintering. J Solid State Electrochem 23:1249–1256. https://doi.org/10.1007/s10008-019-04225-5
35. Hong HYP (1976) Crystal structures and crystal chemistry in the system Na1+xZr2SixP3-xO12. Mater Res Bull 11:173–182. https://doi.org/10.1016/0025-5408(76)90073-8
36. Giarola M, Sanson A, Tietz F et al (2017) Structure and vibrational dynamics of NASICON-Type LiTi2(PO4)3. J Phys Chem C 121:3697–3706. https://doi.org/10.1021/acs.jpcc.6b11067
37. Huang C, Yang G, Yu W, et al (2021) Gallium-substituted Nasicon Na3Zr2Si2PO12 solid electrolytes. J Alloys Compd 855. https://doi.org/10.1016/j.jallcom.2020.157501
38. Guin M, Tietz F (2015) Survey of the transport properties of sodium superionic conductor materials for use in sodium batteries. J Power Sources 273:1056–1064. https://doi.org/10.1016/j.jpowsour.2014.09.137
39. Wang J, He T, Yang X, et al (2023b) Design principles for NASICON super-ionic conductors. Nat Commun 14. https://doi.org/10.1038/s41467-023-40669-0
40. Yin JH, Zhu H, Yu SJ et al (2023) Recent advances of LATP and their NASICON structure as a solid-state electrolyte for lithium-ion batteries. Adv Eng Mater 25:1–17. https://doi.org/10.1002/adem.202300566
41. Zhang Y, Liu H, Xie Z, et al (2023) Progress and perspectives of lithium aluminum germanium phosphate-based solid electrolytes for lithium batteries. Adv Funct Mater 33. https://doi.org/10.1002/adfm.202300973
42. Li S, Zhang SQ, Shen L, et al (2020) Progress and perspective of ceramic/polymer composite solid electrolytes for lithium batteries. Adv Sci 7. https://doi.org/10.1002/advs.201903088
43. Fenton DE, Parker JM, Wright PV (1973) Complexes of alkali metal ions with poly(ethylene oxide). Polymer (Guildf) 14:589. https://doi.org/10.1016/0032-3861(73)90146-8
44. Maurya DK, Dhanusuraman R, Guo Z, Angaiah S (2022) Composite polymer electrolytes: progress, challenges, and future outlook for sodium-ion batteries. Adv Compos Hybrid Mater 5:2651–2674. https://doi.org/10.1007/s42114-021-00412-z
45. Gao H, Zhou W, Park K, Goodenough JB (2016) A sodium-ion battery with a low-cost cross-linked gel-polymer electrolyte. Adv Energy Mater 6. https://doi.org/10.1002/aenm.201600467
46. Tang S, Guo W, Fu Y (2021) Advances in composite polymer electrolytes for lithium batteries and beyond. Adv Energy Mater 11:1–29. https://doi.org/10.1002/aenm.202000802
47. Liang H, Wang L, Wang A, et al (2023) Tailoring practically accessible polymer/inorganic composite electrolytes for all-solid-state lithium metal batteries: a review. Springer. https://doi.org/10.1007/s40820-022-00996-1
48. Yang X, Liu J, Pei N, et al (2023) The critical role of fillers in composite polymer electrolytes for lithium battery. Springer. https://doi.org/10.1007/s40820-023-01051-3
49. Zhang Z, Xu K, Rong X et al (2017) Na3.4Zr1.8Mg0.2Si2PO12 filled poly(ethylene oxide)/Na(CF3SO2)2N as flexible composite polymer electrolyte for solid-state sodium batteries. J Power Sources 372:270–275. https://doi.org/10.1016/j.jpowsour.2017.10.083
50. Dinachandra Singh M, Dalvi A (2021) Ionic transport in NASICON-polymer hybrids: an assessment using X-ray photoelectron spectroscopy. Appl Surf Sci 536:147792. https://doi.org/10.1016/j.apsusc.2020.147792
51. Han F, Zhu Y, He X et al (2016) Electrochemical stability of Li10GeP2S12 and Li7La3Zr2O12 solid electrolytes. Adv Energy Mater 6:1–9. https://doi.org/10.1002/aenm.201501590

52. Cheng L, Wu CH, Jarry A et al (2015) Interrelationships among grain size, surface composition, air stability, and interfacial resistance of al-substituted Li7La3Zr2O12 solid electrolytes. ACS Appl Mater Interfaces 7:17649–17655. https://doi.org/10.1021/acsami.5b02528
53. Gao Y, Wang D, Li YC et al (2018) Salt-based organic-inorganic nanocomposites: towards a stable lithium metal/Li10GeP2S12 solid electrolyte interface. Angew Chem-Int Ed 57:13608–13612. https://doi.org/10.1002/anie.201807304
54. Zheng B, Zhu J, Wang H et al (2018) Stabilizing Li10SnP2S12/Li interface via an in situ formed solid electrolyte interphase layer. ACS Appl Mater Interfaces 10:25473–25482. https://doi.org/10.1021/acsami.8b08860
55. Tatsuma T, Taguchi M, Oyama N (2001) Inhibition effect of covalently cross-linked gel electrolytes on lithium dendrite formation. Electrochim Acta 46:1201–1205. https://doi.org/10.1016/S0013-4686(00)00706-4
56. Chang HJ, Ilott AJ, Trease NM et al (2015) Correlating microstructural lithium metal growth with electrolyte salt depletion in lithium batteries using 7Li MRI. J Am Chem Soc 137:15209–15216. https://doi.org/10.1021/jacs.5b09385
57. Park HE, Hong CH, Yoon WY (2008) The effect of internal resistance on dendritic growth on lithium metal electrodes in the lithium secondary batteries. J Power Sources 178:765–768. https://doi.org/10.1016/j.jpowsour.2007.12.081
58. Cheng XB, Zhang R, Zhao CZ et al (2015) A review of solid electrolyte interphases on lithium metal anode. Adv Sci 3:1–20. https://doi.org/10.1002/advs.201500213
59. Goodenough JB, Hong HY, Kafalas JA (1976) Fast Na^{+}-ion transport in skeleton structures. Mater Res Bull 11:77843. https://doi.org/10.1016/0025-5408(76)90077-5
60. Monroe C, Newman J (2005) The impact of elastic deformation on deposition kinetics at lithium/polymer interfaces. J Electrochem Soc 152:A396. https://doi.org/10.1149/1.1850854
61. Lee B, Paek E, Mitlin D, Lee SW (2019) Sodium metal anodes: emerging solutions to dendrite growth. Chem Rev. https://doi.org/10.1021/acs.chemrev.8b00642
62. Uchida Y, Hasegawa G, Shima K et al (2019) Insights into sodium ion transfer at the Na/NASICON interface improved by uniaxial compression. ACS Appl Energy Mater 2:2913–2920. https://doi.org/10.1021/acsaem.9b00250
63. Fu H, Yin Q, Huang Y et al (2020) Reducing interfacial resistance by Na-SiO2 composite anode for NASICON-based solid-state sodium battery. ACS Mater Lett 2:127–132. https://doi.org/10.1021/acsmaterialslett.9b00442
64. Lu Y, Alonso JA, Yi Q et al (2019) A high-performance monolithic solid-state sodium battery with Ca2+ doped Na3Zr2Si2PO12 electrolyte. Adv Energy Mater 9:1–9. https://doi.org/10.1002/aenm.201901205
65. Zhou W, Li Y, Xin S, Goodenough JB (2017) Rechargeable sodium all-solid-state battery. ACS Cent Sci 3:52–57. https://doi.org/10.1021/acscentsci.6b00321
66. Xu X, Wang S, Wang H et al (2018) Recent progresses in the suppression method based on the growth mechanism of lithium dendrite. J Energy Chem 27:513–527. https://doi.org/10.1016/j.jechem.2017.11.010
67. Yu S, Siegel DJ (2017) Grain boundary contributions to Li-Ion transport in the solid electrolyte Li7La3Zr2O12 (LLZO). Chem Mater 29:9639–9647. https://doi.org/10.1021/acs.chemmater.7b02805
68. Ren Y, Shen Y, Lin Y, Nan CW (2015) Direct observation of lithium dendrites inside garnet-type lithium-ion solid electrolyte. Electrochem commun 57:27–30. https://doi.org/10.1016/j.elecom.2015.05.001
69. Yang J, Liu G, Avdeev M et al (2020) Ultrastable all-solid-state sodium rechargeable batteries. ACS Energy Lett 5:2835–2841. https://doi.org/10.1021/acsenergylett.0c01432
70. Yang J, Wan HL, Zhang ZH et al (2018) NASICON-structured Na3.1Zr1.95Mg0.05Si2PO12 solid electrolyte for solid-state sodium batteries. Rare Met 37:480–487. https://doi.org/10.1007/s12598-018-1020-3
71. Chen Y, Jiang Y, Sen CS et al (2022) Understanding the lithium dendrites growth in garnet-based solid-state lithium metal batteries. J Power Sources 521:230921. https://doi.org/10.1016/j.jpowsour.2021.230921
72. Han F, Westover AS, Yue J et al (2019) High electronic conductivity as the origin of lithium dendrite formation within solid electrolytes. Nat Energy 4:187–196. https://doi.org/10.1038/s41560-018-0312-z
73. Ping W, Wang C, Lin Z et al (2020) Reversible short-circuit behaviors in garnet-based solid-state batteries. Adv Energy Mater 10:1–8. https://doi.org/10.1002/aenm.202000702
74. Schmidt RD, Sakamoto J (2016) In-situ, non-destructive acoustic characterization of solid state electrolyte cells. J Power Sources 324:126–133. https://doi.org/10.1016/j.jpowsour.2016.05.062
75. Aguesse F, Manalastas W, Buannic L et al (2017) Investigating the dendritic growth during full cell cycling of garnet electrolyte in direct contact with Li metal. ACS Appl Mater Interfaces 9:3808–3816. https://doi.org/10.1021/acsami.6b13925
76. Swamy T, Park R, Sheldon BW et al (2018) Lithium metal penetration induced by electrodeposition through solid electrolytes: example in single-crystal Li 6 La 3 ZrTaO 12 garnet. J Electrochem Soc 165:A3648–A3655. https://doi.org/10.1149/2.1391814jes
77. Lu X, Wang Y, Xu X, et al (2023) Polymer-based solid-state electrolytes for high-energy-density lithium-ion batteries—review. Adv Energy Mater 13. https://doi.org/10.1002/aenm.202301746
78. Liang J, Li X, Zhao Y et al (2019) An air-stable and dendrite-free Li anode for highly stable all-solid-state sulfide-based Li batteries. Adv Energy Mater 9:1–10. https://doi.org/10.1002/aenm.201902125
79. Zhang Z, Chen S, Yang J et al (2018) Interface re-engineering of Li10GeP2S12 electrolyte and lithium anode for all-solid-state lithium batteries with ultralong cycle life. ACS Appl Mater Interfaces 10:2556–2565. https://doi.org/10.1021/acsami.7b16176
80. Han F, Yue J, Zhu X, Wang C (2018) Suppressing Li dendrite formation in Li2S-P2S5 solid electrolyte by LiI incorporation. Adv Energy Mater 8:1–6. https://doi.org/10.1002/aenm.201703644
81. Pang Y, Pan J, Yang J et al (2021) Electrolyte/electrode interfaces in all-solid-state lithium batteries: a review. Electrochem Energy Rev 4:169–193. https://doi.org/10.1007/s41918-020-00092-1
82. Sivaraj P, Abhilash KP, Nithyadharseni P, et al (2022) Interfaces in solid-state batteries: challenges and design strategies 193–218. https://doi.org/10.1007/978-3-031-12470-9_7
83. Banerjee A, Wang X, Fang C et al (2020) Interfaces and interphases in all-solid-state batteries with inorganic solid electrolytes. Chem Rev. https://doi.org/10.1021/acs.chemrev.0c00101
84. Eshetu GG, Elia GA, Armand M et al (2020) Electrolytes and interphases in sodium-based rechargeable batteries: recent advances and perspectives. Adv Energy Mater 10:2000093. https://doi.org/10.1002/AENM.202000093
85. Krauskopf T, Richter FH, Zeier WG, Janek J (2020) Physicochemical concepts of the lithium metal anode in solid-state batteries. Chem Rev 120:7745–7794. https://doi.org/10.1021/acs.chemrev.0c00431
86. Xiao Y, Wang Y, Bo SH, et al (2019) Understanding interface stability in solid-state batteries. Nat Rev Mater 5(2):105–126. https://doi.org/10.1038/S41578-019-0157-5
87. Li X (2023) Design principles for interface reaction in all-solid-state batteries. MRS Bull 48:1230–1238. https://doi.org/10.1557/S43577-023-00626-0/METRICS

88. Liu X, Garcia-Mendez R, Lupini AR, et al (2021) Local electronic structure variation resulting in Li 'filament' formation within solid electrolytes. Nature Materials 20(11):1485–1490. https://doi.org/10.1038/s41563-021-01019-x
89. Liang Y, Burton M, Jagger B et al (2024) In situ XPS investigation of the SEI formed on LGPS and LAGP with metallic lithium. Chem Commun 60:12597–12600. https://doi.org/10.1039/D4CC04462B
90. Narayanan S, Ulissi U, Gibson JS et al (2022) Effect of current density on the solid electrolyte interphase formation at the lithium|Li6PS5Cl interface. Nat Commun 13:7237. https://doi.org/10.1038/s41467-022-34855-9
91. Simon FJ, Hanauer M, Henss A et al (2019) Properties of the interphase formed between Argyrodite-Type Li6PS5Cl and polymer-based PEO10:LiTFSI. ACS Appl Mater Interfaces 11:42186–42196. https://doi.org/10.1021/ACSAMI.9B14506/ASSET/IMAGES/LARGE/AM9B14506_0009.JPEG
92. Yang L, Zhang H, Xu H et al (2024) Interfacial catalysis strategy for high-performance solid-state lithium metal batteries. Adv Energy Mater 14:2401829. https://doi.org/10.1002/AENM.202401829
93. Dai J, Yang C, Wang C et al (2018) Interface engineering for garnet-based solid-state lithium-metal batteries: materials, structures, and characterization. Adv Mater 30:1802068. https://doi.org/10.1002/ADMA.201802068
94. Fu KK, Gong Y, Liu B, et al (2017) Toward garnet electrolyte–based Li metal batteries: An ultrathin, highly effective, artificial solid-state electrolyte/metallic Li interface. Sci Adv 3. https://doi.org/10.1126/SCIADV.1601659/SUPPL_FILE/1601659_SM.PDF
95. Pervez SA, Cambaz MA, Thangadurai V, Fichtner M (2019) Interface in solid-state lithium battery: challenges, progress, and outlook. ACS Appl Mater Interfaces 11:22029–22050. https://doi.org/10.1021/ACSAMI.9B02675/ASSET/IMAGES/LARGE/AM-2019-02675T_0011.JPEG
96. Gao Y, Wang D, Li YC et al (2018) Salt-based organic-inorganic nanocomposites: towards a stable lithium metal/Li10GeP2S12 solid electrolyte interface. Angew Chem Int Ed 57:13608–13612. https://doi.org/10.1002/ANIE.201807304
97. Zhu F, Hu X, Xu L et al (2024) Restrained Li|Garnet interface contact deterioration manipulated by lithium modification for solid-state batteries. Adv Funct Mater 34:2314994. https://doi.org/10.1002/ADFM.202314994
98. Wang C, Gong Y, Liu B et al (2017) Conformal, nanoscale ZnO surface modification of garnet-based solid-state electrolyte for lithium metal anodes. Nano Lett 17:565–571. https://doi.org/10.1021/ACS.NANOLETT.6B04695/ASSET/IMAGES/LARGE/NL-2016-046958_0005.JPEG
99. Wang H, Ozkan CS, Zhu H, Li X (2023) Advances in solid-state batteries: Materials, interfaces, characterizations, and devices. MRS Bull 48:1221–1229. https://doi.org/10.1557/S43577-023-00649-7/FIGURES/7
100. Xia W, Xu B, Duan H et al (2017) Reaction mechanisms of lithium garnet pellets in ambient air: the effect of humidity and CO2. J Am Ceram Soc 100:2832–2839. https://doi.org/10.1111/jace.14865
101. Nyman M, Alam TM, McIntyre SK et al (2010) Alternative approach to increasing Li mobility in Li-La-Nb/Ta garnet electrolytes. Chem Mater 22:5401–5410. https://doi.org/10.1021/cm101438x
102. Galven C, Corbel G, Le Berre F, Crosnier-Lopez MP (2016) Instability of the ionic conductor Li6BaLa2B2O12 (B = Nb, Ta): barium exsolution from the garnet network leading to CO2 capture. Inorg Chem 55:12872–12880. https://doi.org/10.1021/acs.inorgchem.6b02238
103. Cheng L, Crumlin EJ, Chen W et al (2014) The origin of high electrolyte-electrode interfacial resistances in lithium cells containing garnet type solid electrolytes. Phys Chem Chem Phys 16:18294–18300. https://doi.org/10.1039/c4cp02921f
104. Li Y, Chen X, Dolocan A et al (2018) Garnet electrolyte with an ultralow interfacial resistance for li-metal batteries. J Am Chem Soc 140:6448–6455. https://doi.org/10.1021/jacs.8b03106
105. Li Y, Xu B, Xu H et al (2017) Hybrid polymer/garnet electrolyte with a small interfacial resistance for lithium-ion batteries. Angewandte Chemie - International Edition 56:753–756. https://doi.org/10.1002/anie.201608924
106. Muramatsu H, Hayashi A, Ohtomo T et al (2011) Structural change of Li2S-P2S5 sulfide solid electrolytes in the atmosphere. Solid State Ion 182:116–119. https://doi.org/10.1016/j.ssi.2010.10.013
107. Hayashi A, Muramatsu H, Ohtomo T et al (2013) Improvement of chemical stability of Li3PS4 glass electrolytes by adding MxOy (M = Fe, Zn, and Bi) nanoparticles. J Mater Chem A Mater 1:6320–6326. https://doi.org/10.1039/c3ta10247e
108. Banerjee A, Park KH, Heo JW et al (2016) Na 3 SbS 4: a solution processable sodium superionic conductor for all-solid-state sodium-ion batteries. Angew Chem 128:9786–9790. https://doi.org/10.1002/ange.201604158
109. Liu L, Xu J, Wang S, et al (2019) Practical evaluation of energy densities for sulfide solid-state batteries. eTransportation 1:100010. https://doi.org/10.1016/j.etran.2019.100010
110. Schnell J, Knörzer H, Imbsweiler AJ, Reinhart G (2020) Solid versus liquid—a bottom-up calculation model to analyze the manufacturing cost of future high-energy batteries. Energy Technology 8. https://doi.org/10.1002/ente.201901237
111. Zhu Y, He X, Mo Y (2015) Origin of outstanding stability in the lithium solid electrolyte materials: insights from thermodynamic analyses based on first-principles calculations. ACS Appl Mater Interfaces 7:23685–23693. https://doi.org/10.1021/acsami.5b07517
112. Schwietert TK, Arszelewska VA, Wang C et al (2020) Clarifying the relationship between redox activity and electrochemical stability in solid electrolytes. Nat Mater 19:428–435. https://doi.org/10.1038/s41563-019-0576-0
113. Calès B, Levasseur A, Fouassier C et al (1977) Conductivite ionique du lithium dans les solutions solides de structure boracite Li4+xB7O12+x/2X (X = Cl, Br) ($0 \leq x \leq 1$). Solid State Commun 24:323–325. https://doi.org/10.1016/0038-1098(77)90219-8
114. Singh MD, Dalvi A Ionic transport in NASICON-polymer hybrids: An assessment using X-ray photoelectron spectroscopy. Appl Surf Sci 536:147792. https://doi.org/10.1016/j.apsusc.2020.147792
115. Xu R, Han F, Ji X et al (2018) Interface engineering of sulfide electrolytes for all-solid-state lithium batteries. Nano Energy 53:958–966. https://doi.org/10.1016/j.nanoen.2018.09.061
116. Zhang Z, Zhang Q, Shi J et al (2017) A self-forming composite electrolyte for solid-state sodium battery with Ultralong cycle life. Adv Energy Mater 7:1–11. https://doi.org/10.1002/aenm.201601196

Resiliency Framework for Overhead Power Distribution Network with Mobile Microgrid

Bonu Ramesh Naidu and Prabodh Bajpai

1 Introduction

Natural disasters, cyberattacks, earthquakes, and other high-impact, low-probability (HILP) events are becoming more frequent and have a major effect on the infrastructure of the electrical system. The overhead power distribution networks (OPDNs) stand out to be the most vulnerable by these events given their vast geographic cover and close proximity to the foreign objects, thereby posing considerable challenges in maintaining uninterrupted service to critical loads (CLs). This led to the development of methodologies for facilitating and improving the performance of OPDNs during HILP events which is broadly addressed as resilient operation. A resilient OPDN must swiftly absorb, adapt, and recover from HILP events to ensure operational normalcy, leading to improved customer satisfaction through continuous electricity supply and minimized safety hazards. The resiliency studies involve long-term planning activities and short-term operational aspects [1]. A methodology that considers both planning and operational activities pertaining to the resilient OPDN is need of the hour.

In contrast to normal operating conditions, prompt and accurate decision-making by the network operator becomes crucial for maintaining OPDN resilience during HILP events. Concurrently, the OPDNs are incorporating emergency backup resources like mobile microgrid (MM) for supporting the network during HILP events. An MM is a truck-mounted or towable microgrid, typically with the utility scale capacity that is used as an emergency backup unit for the OPDNs. Compared to stationary microgrids, their mobility provides operational flexibility to support geographically dispersed loads across an outage area. Developing suitable resiliency framework becomes challenging when considering MM during HILP events as their location is critical for enabling or enhancing the resiliency of OPDNs.

This chapter discusses a novel framework that performs the four major stages of OPDN resiliency, i.e., assessment, quantification, enabling, and enhancement by considering the participation of MMs. The resiliency assessment is performed through metrics that capture two major properties of the OPDN, i.e., physical and topological. The resiliency quantification is performed through Choquet integral and fuzzy measure that facilitates the inclusion of individual importance along with the interaction among various resiliency metrics. Resilience-enabling algorithms for preventive response and operational contingency scenarios are outlined in this work followed by an operational resiliency enhancement algorithm that decides the location of the MM. The framework aids the resilient operation of the OPDN during HILP events by reconfiguring the network with a suitable feasible network considering the participation of MM. The proposed techniques are tested through simulation studies considering the field data of a 415 V OPDN in Kolkata, India.

2 Proposed Resiliency Framework

The suggested resiliency framework is shown in Fig. 1. Considering various physical system and topological metrics for the FNs, *resiliency assessment*, *resiliency quantification*, *resiliency enabling* as a preventive or emergency response, and *resiliency enhancement* based on the operational measures using MM units are the main stages of this framework. The suggested framework considers the physical system metric factor data, bus data, line data, and the damaged lines and nodes following an HILP event as its input.

B. R. Naidu
School of Electrical Engineering and Computer Science, The University of Queensland, Brisbane, Australia
e-mail: r.bonu@uq.edu.au

P. Bajpai (✉)
Department of Sustainable Energy Engineering, Indian Institute of Technology, Kanpur, India
e-mail: pbajpai@iitk.ac.in

S. Narayanan et al. (eds.), *Materials, Devices and Systems for Sustainability*, IITK Directions 8,
https://doi.org/10.1007/978-981-96-7295-0_3

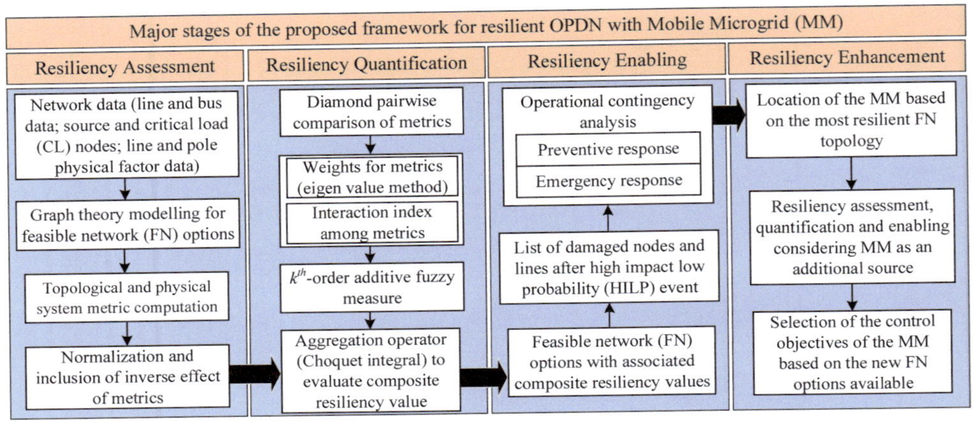

Fig. 1 Resiliency framework for OPDN considering MMs. Adapted and reproduced with permission from [2]

2.1 Graph Modeling of OPDN

In order to analyze resilience and suggest improvements, OPDN modeling is essential. A graph G (N, L) is the OPDN under study with numerous lines and poles, where number of nodes that usually correspond to the poles in the OPDN is represented by N and L is the number of lines. The DFS method [3] is used to obtain all paths between every CL and source pertaining to the network considering all switches as closed. The remaining paths after eliminating paths with meshes and multiple sources are referred to as path combination without loop (PCWL). In order to guarantee the radial operation of the OPDN, various PCWLs are conjoined to create configurations of possible networks (PN), in which each CL is fed from a single source at a time. After then, PNs are grouped according to comparable switch configurations; each group is referred to as a unique network (UN) since it has a distinct switching configuration. In order to keep the OPDN operating radially, the UNs with meshes are also removed. Finally, feasible networks (FNs) that meet the load flow requirements are filtered from UNs.

2.2 Resiliency Assessment of OPDN

A method to quantify the resiliency of a network is vital to evaluate the measures employed to improve resiliency. Assessing resilience aids in determining how well resilience-enhancing techniques are working and allows for any necessary adjustments. The definitions and mathematical formulation of the physical system metrics and topology utilized in this work are provided below [1, 2].

Branch Count Effect The ratio of all connected branches for each PCWL in a PN to all CLs serves as a representation of this. The average value of every PN in the FN is taken into account for each FN.

$$BCE_k = \frac{\sum_{i=1}^{N_{pn}^k}\left(\frac{\sum_{j=1}^{C_n}(\text{Number of edges in a PCWL})}{C_n}\right)}{N_{pn}^k}, \tag{1}$$

where BCE_k is the value for BCE metric for the kth FN, N_{pn}^k is the number of PNs in the kth FN, and C_n is the number of CLs in the PN which is also equal to the number of PCWLs in the PN. A higher BCE requires a bigger part of the network to supply the CLs which may result in a higher probability of failure during HILP events. Hence, the BCE metric and the composite resiliency of FN are correlated through inversely proportional relationship.

Overlapping Branches It is the summation of branches that each pair of PCWLs in a PN has in common. The average value of every PN in the FN is taken into account for each FN.

$$OB_k = \frac{\sum_{i=1}^{N_{pn}^k}(\text{Number of common edges in } i\text{th PN})}{N_{pn}^k}, \tag{2}$$

where OB_k is the value for OB metric for the kth FN. A higher value of OB signifies that most of the CLs are being fed through a few critical and common branches. This increases the single point of failure probability of the OPDN. The OB metric and the composite resiliency of FN are correlated through inversely proportional relationship.

Switching Operations It is defined as the summation of switch state changes—that is, from normally open to closed and from normally closed to open—that are necessary to connect all CLs to a source and generate distinct FNs without any loops.

$$SO_k = \text{Number of switching operations performed to obtain the FN}, \tag{3}$$

where SO_k is the value for SO metric for the kth FN. A switching operation is counted if a sectionalize switch (initially closed) is opened or a tie line (initially open) is closed. It is preferable to achieve the FNs with minimum switching operations. Hence, the SO metric and the composite resiliency of FN are correlated through inversely proportional relationship with a maximum value equal to the number of tie lines and sectionalizers in the network and a minimum value of 0.

Probability of Availability and Penalty Factor To differentiate the kind of source supplying power to the CL, this factor consists of two parts. One is predicated on the source's dependability or likelihood of availability, while the other is predicated on distribution network losses.

$$PoA\&PF_k = \frac{\sum_{i=1}^{N_{pn}^k}(PoA_i \times PF_i)}{N_{pn}^k}, \tag{4}$$

where $PoA\&PF_k$ is the value for $PoA\&PF$ metric for the kth FN, N_{pn}^k is the number of PNs in the kth FN, and PoA_i and PF_i are the values of probability of availability and the penalty factor for ith PN of the FN, respectively. The values of PoA and PF are individually designated a value within the range of 0 and 1. A higher *PoA* value indicates a more reliable source, i.e., the main grid. A higher *PF* indicates fewer losses in the distribution of power, i.e., when using the DG units. The $PoA\&PF_k$ metric and the composite resiliency of FN are correlated through directly proportional relationship with a range of 0 to 1.

Aggregated Central Point Dominance The significance of the node for the specified network topology is captured by this measure. It can be defined mathematically as the mean of the product of each node's degree and betweenness centrality in a given FN.

$$ACPD_k = \frac{\sum_{d=1}^{D_k} \Omega(d) \times C_B(d)}{D_k}, \tag{5}$$

where $ACPD_k$ is the value for ACPD metric for the kth FN, D_k is the number of unique nodes in the kth FN, $\Omega(d)$ is the degree of a particular node in the FN, and $C_B(d)$ is the betweenness centrality of a particular node in the FN. The betweenness centrality is given by the following equation:

$$C_B(d) = \frac{\sum_{j \neq d \neq i} \sigma_{ij}(d)}{\sigma_{ij}}, \tag{6}$$

where σ_{ij} is the total number of shortest paths between nodes i and j, and $\sigma_{ij}(d)$ is the total number of those paths that pass through node d. A higher value of ACPD results in higher importance for any node leading to a higher probability for the single point of failure. Hence, the ACPD metric and the composite resiliency of FN are correlated through inversely proportional relationship.

Line Strength Index The effect of various factors like number of joints, line sag, number of spacers, right-of-way clearance, and per unit length on the strength of lines in a distribution network is captured in this metric. Based on the aforementioned considerations, a weight is allocated to each line in the OPDN based on its physical state. These factors can be added based on the variables that impact an overhead line's performance during an HILP event; they are not all-inclusive. Since lines free of vegetation encroachment are regarded as resilient in this work, this component is given a value of 1 for those lines and 0 otherwise. A line is considered robust if the joint count across its span is less than or equal to 1; in such case, its factor is set to one; otherwise, it is set to zero. In a comparable manner, a line is considered less resilient if its sag exceeds the allowable limit; therefore, the factor value for such line is 0. This factor is given a value of 1 if the number of spacers across a per unit length on the overhead line is equal to or greater than a predetermined value, and 0 otherwise. Additionally, the line weight (Ψ_m) of the network's mth line is calculated while accounting for the contributions of the different LSI factors.

$$\Psi_m = \sum_{i=1}^{l} \left(f_{im} \times W_i^{line} \right), \tag{7}$$

where $W_i line$ denotes the weight of the ith factor chosen by the user to determine the overall weight of the line, $f_i m$ is the assigned value (1 or 0) to the ith factor of the mth line of the network based on the actual data, and l is the number of factors in the LSI metric. The LSI metric is calculated for each FN as the weighted average of the different variables, as indicated below, and the weight of each line in the network will be determined using (7).

$$LSI_k = \frac{\sum_{m=1}^{N_{lk}} \Psi_{mk}}{N_{lk}}, \tag{8}$$

where $Psi_m k$ is the weight of the mth line in the kth FN and $N_l k$ is the number of lines in the kth FN. The FN's composite resiliency value has a directly proportional relation with the LSI metric value.

Pole Strength Index This metric facilitates the capture of the impact of various physical factors that effect the OPDN poles like muffing status, pole sharing, pole corrosion, pole material, guy wire, and pole deflection representing the strength of the poles. Each pole of the network (shown as a node in the graph) is given a weight based on its physical state using the previously mentioned parameters in order to determine the PSI metric for an FN. These are only a few examples and there is always a scope to consider various other additional parameters to capture an electric pole's performance during HILP incidents. A network pole is considered less resilient if it hosts multiple third-party attachments (such as television network cables, telecom wires, etc.), is not designed with mild steel, is not supported by a guy wire, has poor or damaged muffing at the pole base, or if it has deflection and corrosion. All of the PSI metric's factors will be given a score of 1 in these circumstances and 0 otherwise. Further, the weight (Φ_b) of bth node in the network is derived using the following relation:

$$\phi_b = \sum_{p=1}^{r} \left(f_{pb} \times W_p^{node} \right), \quad (9)$$

where f_{pb} is the assigned value to the pth factor of the bth node of the network, W_p^{node} is the weight of the pth factor provided by the user to calculate the overall weight of the node, and r is the number of factors in the PSI metric. Each node's weight in the network is calculated in this manner. The weighted mean of the individual elements, as stated below, is used to calculate the PSI of an FN.

$$PSI_k = \frac{\sum_{b=1}^{N_{nk}} \phi_{bk}}{N_{nk}}, \quad (10)$$

where ϕ_{bk} is the weight of the bth node in the kth FN and N_{nk} is the number of nodes in the kth FN. The FN's composite resiliency score has a directly proportional relation with the PSI metric value.

Resiliency Metric Value Transformations For every FN in the network, metrics of the physical and topological systems are calculated. Due to the fundamental definition of the metric and its mathematical formulation as provided in [1], the magnitude of the various resiliency metrics computed for each given FN varies. Consequently, the composite resiliency score will be dominated by the metric with the widest range of values, which will determine which FN is the most resilient, resulting in an incorrect ranking of the FN possibilities. Additionally, a few metrics are directly proportional to the overall resiliency and a few are inversely proportional. A higher value of inversely proportional metrics should be penalized and given lower priority in the composite resiliency and vice versa. To obviate these effects, a few mathematical transformations (normalization and inclusion of inverse effect) on the individual metric values are proposed in this work. The resiliency metric values are normalized with the highest value among all FNs corresponding to the particular metric. Mathematically it is represented as shown below:

$$\hat{x}_{(i)}(k) = \frac{x_{(i)}}{max\{x_{(i)}(1), x_{(i)}(2), ...x_{(i)}(N_f)\}}, \quad (11)$$

where $\hat{x}_{(i)}$ is the normalized value of ith metric for kth FN on a scale of 0 to 1 and N_f is the number of FNs in the network. The normalized values of the inversely proportional metrics are further modified as per the following equation to account for the inverse effect of metric values:

$$\hat{x}'_{(i)}(k) = 1 - \hat{x}_{(i)}(k), \quad (12)$$

where $\hat{x}'_{(i)}(k)$ is the value of the ith metric for kth FN after considering the effect of inversely proportional relation. The metric is still on a scale of 0 to 1.

2.3 Resiliency Quantification of OPDN

The values of the resiliency metrics for each FN that are on a scale of 0 to 1 emerged out of the resiliency assessment performed in the previous section. The process of quantifying resilience is adding up various measures' values while taking into account each one's significance and how the metrics interact, as shown in Fig. 1. The quantification of resiliency is described as a decision-making process that involves multiple criteria, taking into account the individual significance, overall diversity, and the interactive or correlation nature of the resiliency metrics. The resiliency quantification method is modified in this work by using kth-order additive fuzzy measure and the diamond pairwise comparison (DPC) that accounts for the interactive nature of metrics. The resiliency quantification using the fuzzy measure and DPC involves capturing user input through DPC and generating fuzzy measure for aggregation using the kth-order additive fuzzy measure theory as depicted in Fig. 2.

Analytic Hierarchy Process Analytical hierarchical process (AHP) was originally developed by Prof. Thomas L. Saaty, Distinguished University Professor at the University of Pittsburgh. To summarize, AHP involves paired comparisons to derive ratio scales. The input can be from subjective opinions like preference or from exact measurements like weight, etc. The consistency index is obtained from the principal eigenvalue, and the ratio scales are derived from the principal eigenvectors. The basis for this approach is provided in [4]. All metrics (criteria) are assigned user-defined input weights ϵ (0,9]. Following the determination of pairwise weights, a reciprocal matrix is created, and the primary eigenvector of the reciprocal matrix yields the priority vector, which indicates the relative importance of each criterion.

Diamond Pairwise Comparison As seen in Fig. 3, the DPC involves a diamond shape comprising of vertices representing the relative importance and the interaction between two metrics. It is a graphical interpretation method. Given that i and j are two OPDN metrics, the user's subjective assessment of these metrics can be represented by a point within the

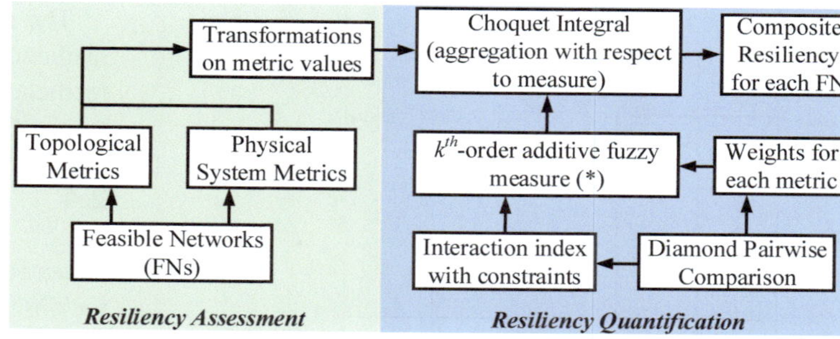

Fig. 2 Composite resiliency quantification for FNs using kth-order additive fuzzy measure. Adapted and reproduced with permission from [2]

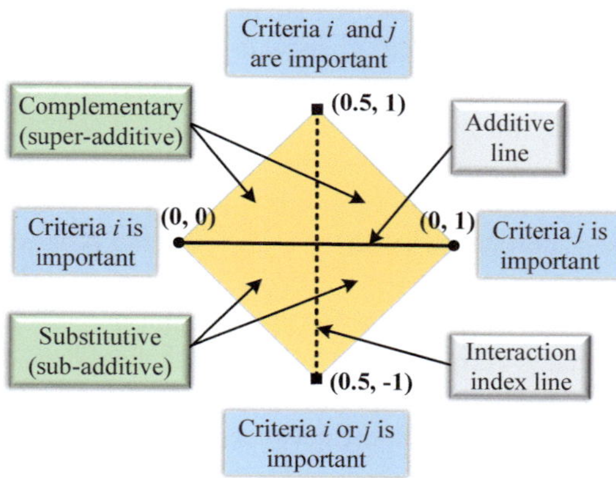

Fig. 3 Diamond pairwise comparison of a pair of metrics. Adapted and reproduced with permission from [2]

diamond shape, i.e., (p,q), where q and p stand for the degree of interaction between metric i and metric j, and the relative importance of metric i in relation to metric j, i.e., $I_{ij}=q$, respectively. As seen in Fig. 3, the type of interaction between the metrics determines the sign of I_{ij}. I_j^{ij} =1-p indicates the relative importance of metric j in relation to metric i. In a multi-criteria decision-making process, the user input about the N metrics available must be captured by a total number of $\frac{N(N-1)}{2}$ pairwise comparisons. Additionally, for all metric combinations, an important ratio matrix (C) of size (N $\times$ N) is created by taking into account that $c_{ij} = \frac{p}{1-p}$ for the upper half off-diagonal elements, 1 for the diagonal elements, and $c_{ji} = \frac{1}{c_{ij}}$ for the lower half off-diagonal elements.

Further, it is necessary to synthesize the desired interaction index and overall importance. The eigenvalues are obtained for the matrix C and the eigenvector pertaining to the largest of the eigenvalues in the matrix is considered as individual importance of the metrics, i.e., $\mu(x_{(i)})$, $\mu(x_{(j)})$, etc. The following linear transformation is used to further translate the interaction degree from the DPC to the appropriate interaction index (I'_{ij}) values, taking into account the transformation's extreme limits [5].

$$max\left(\frac{\mu(x_{(i)}) \times I_{ij}}{(n-1) \times I_i^{ij}}, \frac{\mu(x_{(j)}) \times I_{ij}}{(n-1) \times I_j^{ij}}\right) \leq I'_{ij} \leq 0 \quad (13)$$

$$0 \leq I'_{ij} \leq min\left(\frac{\mu(x_{(i)}) \times I_{ij}}{(n-1) \times I_i^{ij}}, \frac{\mu(x_{(j)}) \times I_{ij}}{(n-1) \times I_j^{ij}}\right). \quad (14)$$

Equations (13) and (14) are used when $I_{ij} \leq 0$ and $I_{ij} \geq 0$, respectively.

kth-Order Additive Fuzzy Measure A trade-off between the fuzzy measure's complexity and modeling ability is made possible by the kth-order additive fuzzy measure, which is a kth-order approximation of the fuzzy measure's polynomial expression [6]. Since it allows for the modeling of interactions between criteria, k=2 is taken into consideration in this work, which appears to be relevant in real-world applications [5]. The Möbius transform (m_μ) representation for second-order additive fuzzy measure is used to change the individual weights and required interaction index of the previously obtained metrics according to the following equations, where X is the set of all metrics.

$$\mu(x_{(i)}) = m_\mu(\{i\}) + \frac{1}{2}\sum_{j \epsilon X \setminus i} m_\mu(\{i, j\}) \quad (15)$$

$$I'_{ij} = m_\mu(\{i, j\}). \quad (16)$$

Additionally, the second-order additive fuzzy measure is calculated by considering the inverse Möbius transform expression provided below:

$$g(A) = \sum_{B \subset A} m_\mu(B) \qquad \forall A \subset X. \quad (17)$$

Choquet Integral for Aggregation Using the idea of measure, Choquet integral can take into consideration a number of interdependent factors. The Choquet integral offers more versatility than other aggregating techniques, such as weighted arithmetic mean, maximum, and minimum. Since the weighted average approach (such as a linear or Lebesgue integral) views the information source as independent or non-interactive, its weighted impacts are seen as additive. However, CIs are appropriate for aggregation when there are interdependencies among the criteria. In the Choquet integral, the

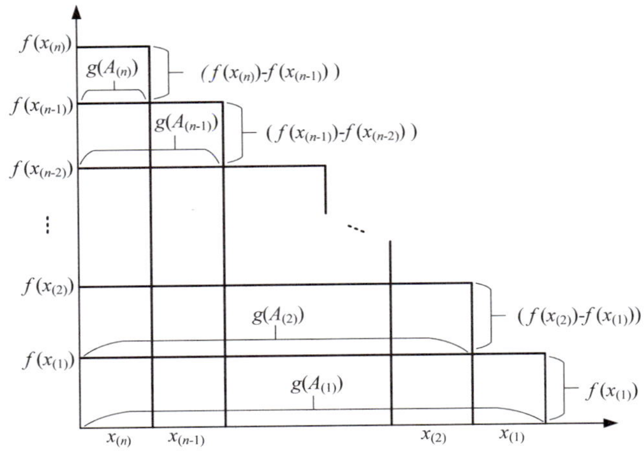

Fig. 4 Graphical representation for computation of CI

idea of measure serves as an index on parameters that aids in interpreting the significance of each criterion.

Let g(X) be a fuzzy measure on X, then Choquet integral of a function f: X→[0,∞] with respect to fuzzy measure g is defined as

$$CI = \sum_{i=1}^{n} (f(x_{(i)}) - f(x_{(i-1)})).g(A_{(i)}), \tag{18}$$

where $f(x_{(0)}) \leq f(x_{(1)}) \leq f(x_{(2)}) \leq f(x_{(3)}) \leq .... \leq f(x_{(n)})$ and A⊂X for i=1,2,3....n, $f(x_{(0)})=0$.

Figure 4 displays the process for calculating the CI. Unlike the traditional integral where the integration is performed by diving the area into multiple rectangles along the horizontal axis, in Choquet integral computation, the area is divided into multiple rectangles along the vertical axis such that the interaction among various factors is considered. Accordingly, the individual areas are given by the following equations:

$$\begin{aligned} Area_1 &= f(x_{(1)}).g(A_{(1)}) \\ Area_2 &= (f(x_{(2)}) - f(x_{(1)})).g(A_{(2)}) \\ Area_3 &= (f(x_{(3)}) - f(x_{(2)})).g(A_{(3)}) \\ Area_n &= (f(x_{(n)}) - f(x_{(n-1)})).g(A_{(n)}), \end{aligned} \tag{19}$$

where

$$\begin{aligned} g(A_{(1)}) &= g(x_{(1)}, x_{(2)}, x_{(3)}...x_{(n)}) \\ g(A_{(2)}) &= g(x_{(2)}, x_{(3)}...x_{(n)}) \\ g(A_{(3)}) &= g(x_{(3)}...x_{(n)}) \\ g(A_{(n)}) &= g(x_{(n)}). \end{aligned} \tag{20}$$

The variables in the above equation are computed using (17) for kth-order additive fuzzy measure. The Choquet integral is then computed using the following equation:

$$CI = \int Fdg = Area_1 + Area_2 + Area_3 + ... + Area_n. \tag{21}$$

The ranking of the FNs on the basis of their composite resiliency value is performed by computing the composite resiliency values using CI.

2.4 Resiliency Enabling of OPDN

Instead of emphasizing an operation with highest resiliency in the absence of an HILP event, the network operator stresses economical, dependable, and secure operation. As a result, under typical operational conditions, the network design with the highest resilience based on the resiliency value might not be deployed in the field. Using hit-and-trial techniques to restore the network's supply to CLs during and after an HILP incident may make matters worse due to the active infrastructure that is available and the strict response times. The algorithm presented in Fig. 5 gives the network operator an organized approach to select the most resilient FN as an emergency response after an HILP event or as a preventive response before an impending HILP event using the network infrastructure that is currently in place. Bus data, load data, source nodes, CL nodes, and LSI and PSI metric factor data—which are often accessible with the network operator beforehand—are among the input data needed for the method. Additionally, the list of nodes and lines lost following the HILP incident, as well as the base case data—PNs, unmeshed UNs, FNs, and their corresponding resilience values—must be entered for the whole network under normal operating conditions.

Preventive Response Considering the entire network during normal operation (without any HILP event), a list of FNs with their resiliency value computed by the proposed assessment and quantification techniques is produced as the base case data. As a preventive response to the prediction of any HILP occurrence from a meteorological forecast, the FN with the highest resiliency value can be selected from this list and activated with the proper switching configuration.

Emergency Response Despite precautionary measures, the spatiotemporal nature of the majority of HILP events can lead to the failure of multiple lines and/or nodes in the network. Due to damaged lines and poles, the OPDN may ultimately dissolve into several sub-networks, each of which contains a subset of the OPDN's sources and CLs. Figure 5 illustrates how the algorithm identifies the sub-networks, as well as their sources and CLs information, and then evaluates each sub-network separately. The algorithm selects a random node from the network and begins constructing the sub-network by obtaining the connected nodes and associated lines when emergency response is initiated by entering the nodes and lines corresponding to the damaged poles and lines in the network. Once the first sub-network has been obtained, the sub-network's lines and nodes are stored independently. The algorithm then repeats the process by selecting a random node

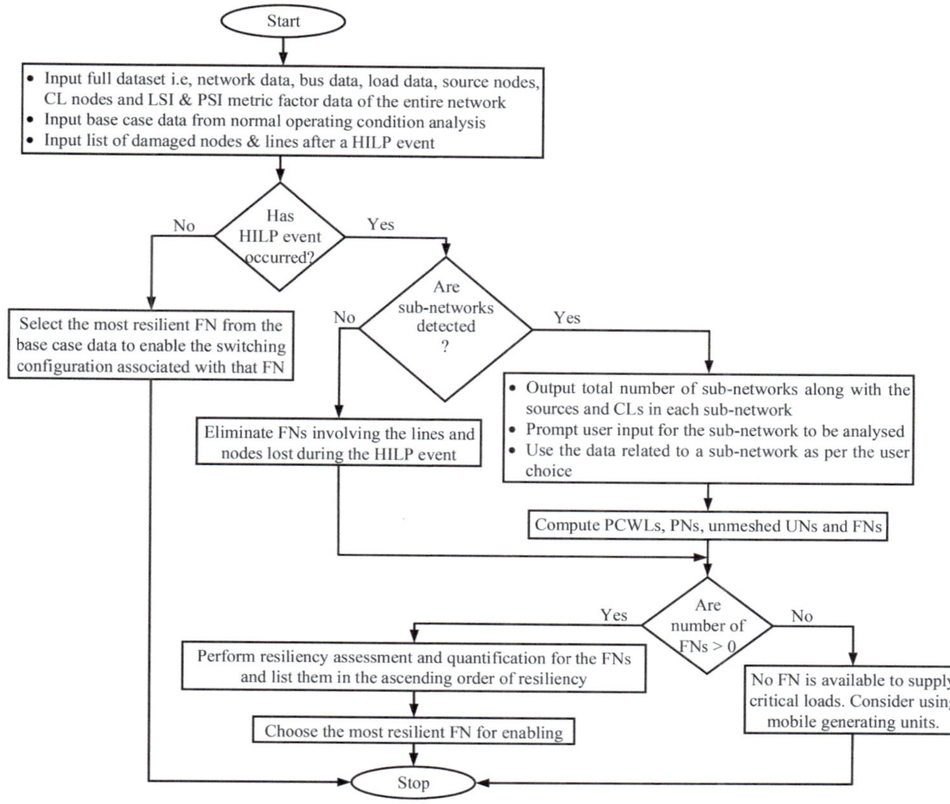

Fig. 5 Algorithm for enabling resilient operation of an OPDN under the operational scenario emerging from a contingency event. Adapted and reproduced with permission from [2]

from the remaining nodes until every node in the active network is connected to a specific sub-network.

After providing information about the possible sub-networks created as a result of the HILP event, the algorithm prompts the user to choose one to be examined. Additionally, a sub-network under examination undergoes resiliency assessment and measurement based on user feedback. If the HILP event did not warrant the establishment of any sub-networks, the FNs from the base case data are selected and filtered to eliminate FNs involving the damaged nodes and lines. The algorithm then eliminates FNs that traverse through damaged lines and damaged nodes in order to select the most resilient FN for the emergency response.

2.5 Operational Resiliency Enhancement

Two significant aspects are critical for the effective participation of MM during HILP events, i.e., the MM's location and the control strategy for MM to transit through different modes that arise during an HILP event. The resiliency enhancement strategy proposed in this work is depicted in Fig. 1. The location of the MM can be decided to either create a more resilient FN option using the MM or to sustain the existing most resilient FN topology available. The first approach requires a repetitive analysis involving resiliency assessment and quantification with MM connected to various nodes of the OPDN to generate a more resilient FN options (if any). This approach seems feasible for OPDNs with less number of nodes/lines and becomes increasingly complex for bigger networks. The second approach indeed considers the most resilient FN topology obtained from the resiliency enabling stage and connects the MM in parallel to the source node of the FN. With this approach and with an appropriate control strategy for the MM, the most resilient topology can be sustained despite the unavailability of the source node feeding the FN. Additionally, grid-supporting features like voltage support during normal operation and fault conditions can be provided by the MM to improve the power quality to the consumers [7]. With the introduction of MM for resiliency enhancement, the location of the MM may impact the direction of the power flow in the lines of the OPDN (if the MM is allowed to connect at any node in the network). However, in this work, the location of the MM is usually in parallel to one of the source nodes available in the active sub-network which takes care of protection against any possible reverse current flow. Therefore, the protection settings are not impacted by the MM operation.

The immediate step after deciding the location of the MM is to find out its capability in sustaining the FN structure as an electrical island in the event of isolation of the existing FN source. Hence, the resiliency assessment, quantification, and enabling are performed with MM connected to the network at the proposed location to identify if the most resilient FN is still feasible while being fed from the MM. The MM can form the island if FN can be sustained as an electrical island by the MM, else, the role of MM is limited to grid-following operation to support the network as a real/reactive power source/sink and to perform grid-supporting activities. This aids in deciding the control objectives of the MM.

3 System Under Study

The CESC Ltd., Kolkata, India's Nayabad (N) O/T low-voltage OPDN is used to simulate and analyze the resilience framework. MATLAB software is used to carry out the complete analysis. The network under investigation is a 415 V OPDN of CESC Ltd. that is primarily used to supply residential loads. It is connected in a ring-main configuration but functions radially from the corresponding distribution transformer. Figure[1] shows the network's single-line diagram with various sources and CLs. This network uses 150 sqmm LT ABC type, ACSR Racoon, and ACSR ferret wires for the pole-to-pole connection. Approximately 20% of the network is made up of underground XLPE cables. Overhead poles are made of mild steel and reinforced cement concrete.

Figure 6b presents the results of the graph modeling. There are 137 nodes and 140 edges in the network graph overall. Five CLs are taken into consideration at nodes 7, 18, 41, 100, and 134, while the node 115 is representing the upstream network (main grid). Six sources from additional distribution transformers are represented at nodes 1, 40, 112, 116, 135, and 136 in this network model. For emergency supply during HILP scenarios, three mobile diesel generator units that have different capacity (63, 250, and 500 kVA) might be regarded as distributed generation units. These mobile DG units are now connected to node 137, which is connected to node 19 via a tie-line switch. The DG unit is considered to have a power factor of 0.85, the apparent power limit for the main grid source is 400 kVA, and the remaining distribution transformer sources are expected to be 125 kVA. To determine the FNs, the load flow analysis takes these constraints into account. There are 41 sectionalizing switches and 11 tie lines in all. The GIS portal provided the number of joints, bus data, number of spacers, line data, line length, pole deflection status, vegetation encroachment, sag, muffing status, third-party attachments on the pole, pole material, and presence of guy wire for providing physical support to the pole used in this study. The network considered for resiliency quantification during normal operating conditions and contingency scenarios is described below.

[1] DTR stands for distribution transformer.

- **Normal Operating Condition:** The entire network under study is considered with only one source, i.e., the main grid pertaining to the network. All tie lines and sectionalizing switches in the network are included, but only CLs are considered for generating the FNs.
- **Contingency Scenario:** Only a part of the network under study (the active network available after an HILP event) is considered along with multiple sources like DG units and other available distribution transformers. All tie lines and sectionalizing switches pertaining to the active network are considered, but only CLs are considered for generating the FNs.

The simulation study is predicated on the following assumptions: (a) all DGs are modeled as PV buses; (b) the network topology can be reconfigured using only tie lines and sectionalizers; (c) the network operator has complete control over opening and closing all switches in the network; (d) the network operator is aware of the list of available poles and lines in the network both during and after an HILP event; (e) the load is balanced for load flow study; and (f) the voltage limits for load flow convergence are assumed to be 1 $\pm$ 0.06 pu unless otherwise specified.

3.1 User Input for the Resiliency Framework

The PoA and PF for the main grid are assumed to be 0.98 and 0.9 in this analysis, while the values for the remaining sources are assumed to be 0.95 and 1, respectively. According to the OPDN GIS portal, the spacer requirement for the OPDN under study is one spacer for every five meters of line length. The CESC Ltd. GIS portal provides the line and bus data of the network being studied as well as the physical system metric factor data. Tables 1 and 2 include the pairwise comparison matrices for the LSI and PSI metrics. AHP is used to determine each metric's unique weights. The LSI factors' weights (W_i^{line}) for the LSI factors in this study are, respectively, 0.5558, 0.1364, 0.0489, and 0.2589. For the PSI factors in this study, the corresponding weights (W_p^{node}) are 0.0672, 0.1235, 0.2330, 0.5120, 0.0380, and 0.0263.

3.2 Computation of kth-Order Additive Fuzzy Measure

The relative importance and interaction index of each pair of metrics obtained through DPC based on the user preference is tabulated in Table 4. The individual importance of metrics is taken similar to the λ-fuzzy measure. The same is represented through I_i^{ij} and I_j^{ij} in Table 4. Because the factors that

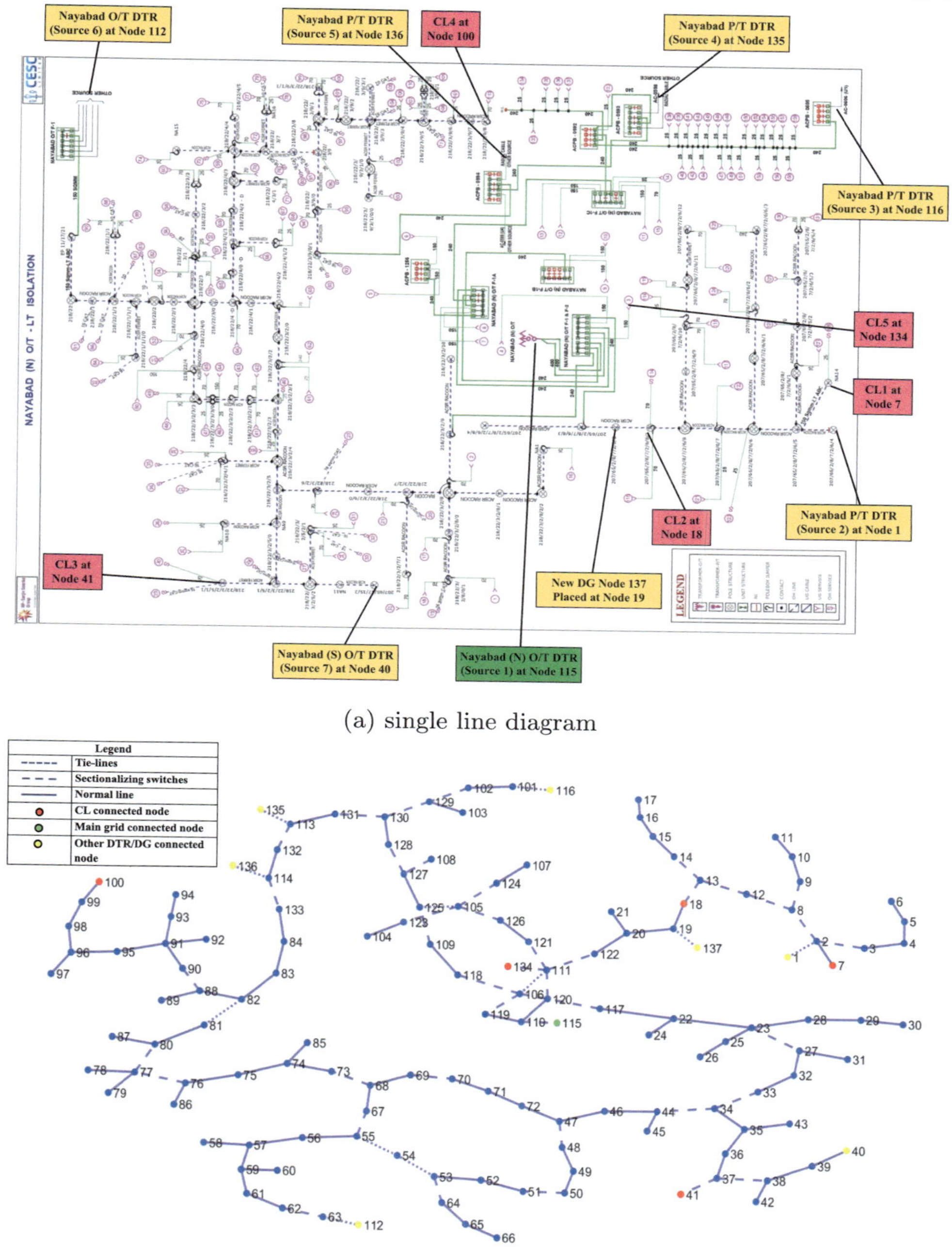

(a) single line diagram

(b) graph plot

Fig. 6 Nayabad (N) O/T network details. Adapted and reproduced with permission from [2]

define the LSI metric partially capture the resiliency status of the PSI metric, the interaction among the physical system metrics is viewed as sub-additive. For instance, vegetation encroachment on lines results in excessive weights on the lines, which in turn causes pole bending, and vice versa. Likewise, it is assumed that the interaction between the topological metrics is sub-additive. Nonetheless, the super-additive interaction between the metrics of the physical and topological systems is taken into account. I_{ij} in Table 4 represents the same. Moreover, (22) provides the importance ratio matrix (C). This matrix is same as the pairwise comparison matrix for λ-fuzzy measure given in Table 3. Hence, the individual weights or the overall importance/weights are obtained through eigenvalue analysis like λ-fuzzy measure. The overall importance/weights are $\mu(X) = 0.0381, 0.0381, 0.0279, 0.105, 0.1868, 0.302, 0.302$ for BCE, OB, SO, PoA&PF, ACPD, LSI, and PSI, respectively, as obtained from matrix C. The desired interaction index (I'_{ij}) obtained from Eqs. (13) and (14) is tabulated in Table 4. Finally, the kth-order additive fuzzy measure is computed.

Table 1 Pairwise comparison matrix for factors in LSI metric

	RoW clearance	No. of joints	Line Sag	No. of spacers
RoW clearance	1/1	5/1	7/1	3/1
No. of joints	1/5	1/1	5/1	1/3
Line Sag	1/7	1/5	1/1	1/5
No. of Spacers	1/3	3/1	5/1	1/1

Table 2 Pairwise comparison matrix for factors in PSI metric

	Sharing	Deflection	Corrosion	Muffing	Guy wire	Material
Sharing	1/1	1/3	1/5	1/5	3/1	3/1
Deflection	3/1	1/1	1/3	1/5	5/1	5/1
Corrosion	5/1	3/1	1/1	1/3	7/1	7/1
Muffing	5/1	3/1	7/1	1/1	9/1	9/1
Guy wire	1/3	1/5	1/7	1/9	1/1	3/1
Material	1/3	1/5	1/7	1/9	1/3	1/1

Table 3 Pairwise comparison matrix for the resiliency metrics

	BCE	OB	SO	PoA&PF	ACPD	LSI	LSI
BCE	1/1	1/1	3/1	1/5	1/6	1/7	1/7
OB	1/1	1/1	3/1	1/5	1/6	1/7	1/7
SO	1/3	1/3	1/1	1/6	1/7	1/5	1/5
PoA&PF	5/1	5/1	6/1	1/1	1/4	1/5	1/5
ACPD	6/1	6/1	7/1	4/1	1/1	1/3	1/3
LSI	7/1	7/1	5/1	5/1	3/1	1/1	1/1
PSI	7/1	7/1	5/1	5/1	3/1	1/1	1/1

$$C = \begin{bmatrix} & \textbf{BCE} & \textbf{OB} & \textbf{SO} & \textbf{PoA\&PF} & \textbf{ACPD} & \textbf{LSI} & \textbf{PSI} \\ \textbf{BCE} & 1 & 1 & 3 & 0.2 & 0.167 & 0.143 & 0.143 \\ \textbf{OB} & 1 & 1 & 3 & 0.2 & 0.167 & 0.143 & 0.143 \\ \textbf{SO} & 0.333 & 0.333 & 1 & 0.167 & 0.143 & 0.2 & 0.2 \\ \textbf{PoA\&PF} & 5 & 5 & 6 & 1 & 0.25 & 0.2 & 0.2 \\ \textbf{ACPD} & 6 & 6 & 7 & 4 & 1 & 0.333 & 0.333 \\ \textbf{LSI} & 7 & 7 & 5 & 5 & 3 & 1 & 1 \\ \textbf{PSI} & 7 & 7 & 5 & 5 & 3 & 1 & 1 \end{bmatrix} \tag{22}$$

4 Simulation Case Studies

This section presents three simulation case studies to validate the proposed framework. The first case study is performed considering normal operating condition of the network. The second case study considers a contingency scenario with multiple active sub-networks. The third case study considers another contingency scenario and the resiliency enhancement algorithm with MM is tested. In the first two cases, the MM (node 137) is connected at node 19 through a tie line whereas for the case study-3, its location is obtained based on the resiliency enhancement algorithm.

Case Study-1: Normal Operating Condition The methodology proposed in Sect. 2.3 is used for quantifying the resiliency of the network under normal operating conditions. PCWLs are obtained between the source node (115) and the CL nodes (7, 18, 41, 100, and 134). A total of 36 PCWLs are available as listed in Table 5. Further, PNs are formed from the available PCWLs such that all CLs are supplied from a single source at a time. A total of 13,750 PNs are available. It is most likely that the switching configuration (switches traversed in the path) of one or more PNs may be identical, thus leading to a possibility of combining the same into a single UN, e.g., the switches involved in PN161 are tie-line switches [53-54, 54-55, 81-82, 106-111], sectionalizing switches [2-8, 8-12, 12-13, 13-18, 23-27, 33-34, 37-41, 34-44, 47-48, 50-51, 55-67, 69-70, 68-73, 76-77, 77-80, 88-90, 110-115, 106-117, 106-118, 106-119, 111-122, 125-109, 128-127, 130-131, 114-133, 113-132, 111-134], and all are closed.

PN206 contains the same switches in the PN configuration with all the switches closed. Both of them are combined to form UN37 as per the proposed methodology. Hence, 103 UNs are formed from the 13,750 PNs of the network. The UNs thus obtained are fed from only one source but may have meshes in the PNs by virtue of the combination of PCWLs to PNs or PNs to UNs. Hence, the UNs are further filtered to eliminate UNs involving meshes. The final list of meshed and unmeshed UNs is presented in Table 6. It is evident that, out of the 103 UNs, only 32 UNs emerged as unmeshed UNs. The meshed UNs are discarded and the load flow algorithm is performed on the unmeshed UNs using the backward-forward sweep method to obtain the FNs. Accordingly, 14 out of the 32 unmeshed UNs qualified as FNs. The list of FNs and their associated UN numbers are presented in Table 6. The resiliency assessment and quantification is performed on the FNs to rank them in the order of their composite resiliency values in Table 8. An illustrative computation for FN1 is given in Table 7. The most resilient FN, i.e., FN7 is shown in Fig. 7.

Preventive Response: The base case data for the preventive response analysis consists of the PNs, unmeshed UNs, and FNs from the analysis of normal operating conditions. The most resilient FN, FN 7, with a resiliency value of 74.78%, is activated by the necessary switching configuration when an HILP event is anticipated based on the weather forecast. By taking this step, the network operator can help the network better withstand the effects of HILP events and maintain power for the CLs.

Case Study-2: Contingency Scenario Multiple pole and line failures in the OPDN during HILP occurrences are highly expected, even with the network operator's preventative response. In Fig. 8, ten poles (represented by nodes 22,

Table 4 User input based on DPC for computing the fuzzy measure. Adapted and reproduced with permission from [2]

i vs j	I_i^{ij}	I_j^{ij}	I_{ij}	c_{ij} and c_{ji}	I'_{ij}
BCE vs. OB	0.500	0.500	−0.200	1.000	−0.0025
BCE vs. PoA&PF	0.167	0.833	−0.067	0.200	−0.0025
BCE vs. SO	0.750	0.250	−0.100	3.000	−0.0019
OB vs. SO	0.750	0.250	−0.100	3.000	−0.0019
BCE vs. LSI	0.125	0.875	0.250	0.143	0.0144
BCE vs. ACPD	0.143	0.857	−0.286	0.167	−0.0127
OB vs. PoA&PF	0.167	0.833	−0.067	0.200	−0.0025
BCE vs. PSI	0.125	0.875	0.250	0.143	0.0144
OB vs. ACPD	0.143	0.857	−0.057	0.167	−0.0025
SO vs. PoA&PF	0.143	0.857	−0.057	0.167	−0.0019
ACPD vs. PSI	0.250	0.750	0.500	0.333	0.0623
SO vs. ACPD	0.125	0.875	−0.050	0.143	−0.0019
OB vs. PSI	0.125	0.875	0.250	0.143	0.0144
PoA&PF vs. PSI	0.167	0.833	0.330	0.200	0.0346
SO vs. PSI	0.167	0.833	0.330	0.200	0.0199
SO vs. LSI	0.167	0.833	0.330	0.200	0.0199
OB vs. LSI	0.125	0.875	0.250	0.143	0.0144
PoA&PF vs. LSI	0.167	0.833	0.330	0.200	0.0346
LSI vs. PSI	0.500	0.500	−0.200	1.000	−0.0201
ACPD vs. LSI	0.250	0.750	0.500	0.333	0.0623
PoA&PF vs. ACPD	0.200	0.800	−0.080	0.250	−0.007

Table 5 PCWLs for the individual CLs and the overall PNs from the source at node 115

	CL_1 at node 7	CL_2 at node 18	CL_3 at node 41	CL_4 at node 100	CL_5 at node 134	Total PCWLs	Total PNs
From node 115	5	5	11	10	5	5+5+11+10+5 = 36	5×5×11×10×5 = 13750

Table 6 Meshed, unmeshed UNs, and FNs

Condition	Count	UN number
Meshed UNs (discarded)	71	2 3 4 5 6 7 8 9 10 11 12 14 15 16 17 18 19 20 22 23 25 26 28 29 31 32 33 34 35 36 37 38 40 41 42 44 46 47 48 49 50 51 52 54 55 56 58 60 62 64 66 71 72 74 79 80 82 83 85 86 87 88 90 91 92 96 97 98 99 101 102
Unmeshed UNs (to be used further)	32	1 13 21 24 27 30 39 43 45 53 57 59 61 63 65 67 68 69 70 73 75 76 77 78 81 84 89 93 94 95 100 103
UNs qualified as FNs	14	13(FN1), 21(FN2), 30(FN3), 43(FN4), 51(FN5), 65(FN6), 69(FN7), 70(FN8), 75(FN9), 78(FN10), 81(FN11), 89(FN12), 95(FN13), 103(FN14)

23, 24, 25, 26, 28, 29, 30, 83, 84) and the lines that connect with them are surrounded by red dotted lines, signifying damaged nodes, in order to investigate a contingency scenario and emergency response to it. As a result, the network is divided into two active sub-networks, each of which has a subset of certain CLs and sources and is surrounded by green dotted lines in Fig. 8.

– **Case 2A: Resiliency Quantification for Active Sub-network-1:**

This sub-network contains multiple sources and CLs. PCWLs are obtained between the source nodes[2] (115, 116, 1, 135 and 136) and the CL nodes (7, 18, and 134). A total of 39 PCWLs are available. Further, PNs are formed from the available PCWLs such that all CLs are supplied from a single source at a time. A total of 109 PNs are available during the contingency analysis (with all sources and CLs available in the active sub-network-1) against the 13,750 PNs available (with only the main grid source and all CLs of the entire network) under a normal operation scenario. A total of 13 unmeshed UNs are formed from the 109 PNs of the network after combining PNs based on similar switch configurations and eliminating UNs with meshes. The load flow analysis is performed on the unmeshed UNs using backward-forward sweep method to obtain the FNs. Accordingly, 13 unmeshed UNs qualified as FNs. The list of FNs and their associated UN numbers are presented in Table 9. The resiliency assessment and quantification

[2] The DG unit at node 137 is considered in case study-3 and case study-4 only for resiliency enhancement studies.

Table 7 Composite resiliency computation for FN1 using fuzzy measures and Choquet integral

Metrics	BCE(1)	OB(2)	SO(3)	PoA & PF(4)	ACPD(5)	LSI(6)	PSI(7)
Values	0.132	0.322	0.042	1	0.041	0.9933	0.9941
Order	PoA & PF(4) >PSI (7)>LSI (6)>OB (2)>BCE (1)> SO(3)>ACPD (5)						
Fuzzy measures			*k*th-order fuzzy measure				
g(4)			0.0773				
g (4,7)			0.3513				
g(4,6,7)			0.6051				
g (2,4,6,7)			0.6598				
g(1,2,4,6,7)			0.7171				
g(1,2,3,4,6,7)			0.763				
g(1,2,3,5,4,6,7)			1				
Choquet integral Computation for FN1 with *k*th-order additive measure							
CI = [g(1,2,3,4,5,6,7) × (0.041-0) + g(1,2,3,4,6,7) × (0.042-0.041) + g (1,2,4,6,7) × (0.132-0.042) + g(2,4,6,7) × (0.322-0.132) + g(4,6,7) × (0.993-0.322) + g(4,7) × (0.994-0.993) + g(4) × (1-0.994)] = [(1×0.041) + ((0.763×0.001) + (0.7171× 0.09) + (0.6598×0.19) + (0.6051×0.671) + (0.3513×0.001) + (0.0773×0.006)]×100 = 63.85%							

Table 8 FNs with the metric values in the ascending order of composite resiliency for case study-1

	Topological					Physical		
FNs	BCE	OB	SO	PoA& PF	ACPD	LSI	PSI	Composite resiliency (%)
6	0.00	0.00	0.17	1.00	0.02	1.00	1.00	61.75
1	0.13	0.32	0.04	1.00	0.04	0.99	0.99	63.85
9	0.20	0.59	0.17	1.00	0.00	0.99	0.99	65.28
4	0.17	0.41	0.08	1.00	0.07	0.99	0.99	65.30
12	0.18	0.44	0.08	1.00	0.07	0.99	0.99	65.35
14	0.18	0.44	0.08	1.00	0.07	0.99	0.99	65.35
5	0.22	0.26	0.13	1.00	0.32	1.00	0.99	71.34
2	0.37	0.54	0.00	1.00	0.35	0.99	0.98	73.18
13	0.34	0.56	0.13	1.00	0.32	1.00	0.99	73.40
8	0.41	0.70	0.08	1.00	0.29	0.99	0.99	73.44
11	0.40	0.61	0.00	1.00	0.34	0.99	0.98	73.61
3	0.39	0.57	0.04	1.00	0.38	0.99	0.98	74.31
10	0.40	0.60	0.04	1.00	0.38	0.99	0.98	74.52
7	0.42	0.75	0.13	1.00	0.33	0.99	0.98	74.78

for the sub-network-1 are performed and the FNs along with their metric values are listed in the ascending order of their composite resiliency values in Table 10. The network operator can enable the switching configuration corresponding to the most resilient FN option, i.e., FN3 having 79.17% resiliency for supplying the CLs in the most resilient manner with the available active infrastructure in the sub-network.

- **Case 2B: Resiliency Quantification for Active Sub-network-2:** This sub-network contains multiple sources and CLs. PCWLs are obtained between the source nodes (112 and 40) and the CL nodes (41 and 100). The available six PCWLs led to the formation of six PNs in this sub-network. A total of five unmeshed UNs are formed from these PNs. The load flow analysis is performed on the unmeshed UNs using backward-forward sweep method to obtain the FNs. Voltage limits considered are 1 ± 0.9 pu for this sub-network.[3] Two unmeshed UNs were therefore eligible to be FNs. In Table 11, the FNs and their metric values are listed in ascending order of their composite resiliency values following the completion of the sub-network-2's resiliency assessment and quantification. In order to offer the CLs with the most resilient active infrastructure in the sub-network-2, the network operator can activate the switching configuration that corresponds to

[3] There are no FNs obtained with less than 1 ± 9% voltage limits.

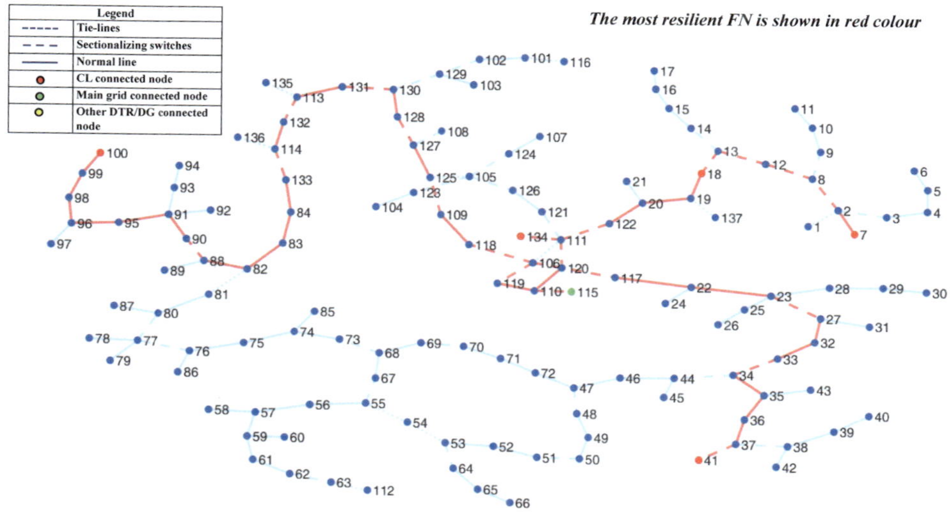

Fig. 7 Graph plot of the test network with FN7 (UN69) highlighted in red. Adapted and reproduced with permission from [2]

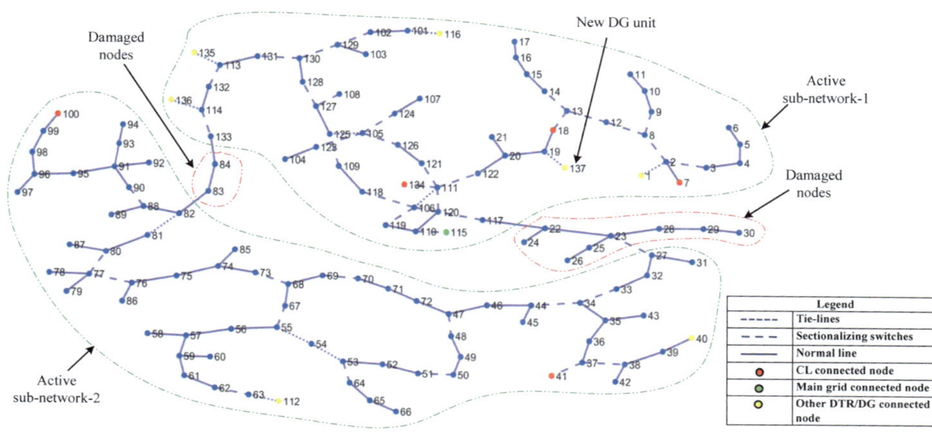

Fig. 8 Graph plot with active sub-networks for contingency analysis. Adapted and reproduced with permission from [2]

Table 9 FNs along with associated UNs of active sub-network-1

Condition	Count	UN number
UNs qualified as FNs	13	1(FN1), 5(FN2), 7(FN3), 8(FN4), 9(FN5), 13(FN6), 15(FN7), 16(FN8), 20(FN9), 22(FN10), 23(FN11), 27(FN12)

the most resilient FN option, FN2, which has 64.97% resiliency.

Case Study-3: Mobile Microgrid Placement Referring to Fig. 9, an operational contingency scenario where 12 poles (represented by nodes 22, 23, 24, 25, 26, 28, 29, 30, 109, 118, 121, and 126) and their associated lines highlighted by the red shaded area is considered as damaged and the emergency response to this event is studied. Two active sub-networks with a subset of sources and CLs were the outcome of this. While the sub-network-2 has 94 lines and 94 nodes with the sources at nodes 40, 112, 116, 135, and 136 and the CLs at nodes 41 and 100, the sub-network-1 has 30 lines and 30 nodes with the source nodes at 1 and 115 and the CLs at nodes 7, 18, and 134.

In this study, the sub-network-1 is taken into consideration for analysis. After computation, three FNs—FN1, FN2, and FN3—are derived from nine PNs and three unmeshed UNs. The FNs and their metric values are listed in Table 12 in ascending order of their composite resiliency values, following the completion of the sub-network-1's resiliency evaluation and quantification. The most resilient FN option in sub-network-1 for feeding the CLs is the FN2 with a source at node 115. Figure 10a shows the graph plot of the FN structure. Therefore, the MM represented by node 137 is connected to node 110 through a tie line to feed the FN2 structure in the absence of a source at 115 as shown in Fig. 10b. Further, the resiliency assessment and quantification of sub-network-1 are performed considering MM as an additional source and the FN options corresponding to MM are tabulated in Table 12 and highlighted in blue color. The additional FN options confirm

Table 10 FNs arranged in the ascending order of their composite resiliency values for sub-network-1 along with their individual metric values. Adapted and reproduced with permission from [2]

FNs	Topological				Physical			
	BCE	OB	SO	PoA& PF	ACPD	LSI	PSI	Composite resiliency (%)
FN13	0.00	0.00	0.33	1.00	0.00	1.00	1.00	62.16
FN7	0.05	0.06	0.33	1.00	0.08	1.00	0.99	64.42
FN10	0.10	0.12	0.29	1.00	0.16	0.99	0.99	66.33
FN12	0.15	0.18	0.19	1.00	0.23	0.98	0.98	67.88
FN11	0.15	0.18	0.19	1.00	0.23	0.98	0.98	67.88
FN6	0.20	0.24	0.19	1.00	0.30	0.98	0.97	69.81
FN5	0.20	0.24	0.19	1.00	0.30	0.98	0.97	69.81
FN2	0.25	0.31	0.29	0.93	0.37	0.97	0.96	71.24
FN9	0.25	0.31	0.14	1.00	0.37	0.97	0.96	71.41
FN8	0.25	0.31	0.14	1.00	0.37	0.97	0.96	71.41
FN4	0.71	0.86	0.00	1.00	0.81	0.80	0.75	76.23
FN1	0.56	0.67	0.10	0.93	0.70	0.91	0.87	78.33
FN3	0.61	0.73	0.14	0.93	0.74	0.89	0.85	79.17

Table 11 FNs in the ascending order of composite resiliency for sub-network-2 with their metric values. Adapted and reproduced with permission from [2]

FNs	Topological				Physical			
	BCE	OB	SO	PoA& PF	ACPD	LSI	PSI	Composite resiliency (%)
FN1	0.00	0.22	0.00	1.00	0.00	1.00	1.00	61.72
FN2	0.02	0.00	0.13	1.00	0.17	0.98	1.00	64.97

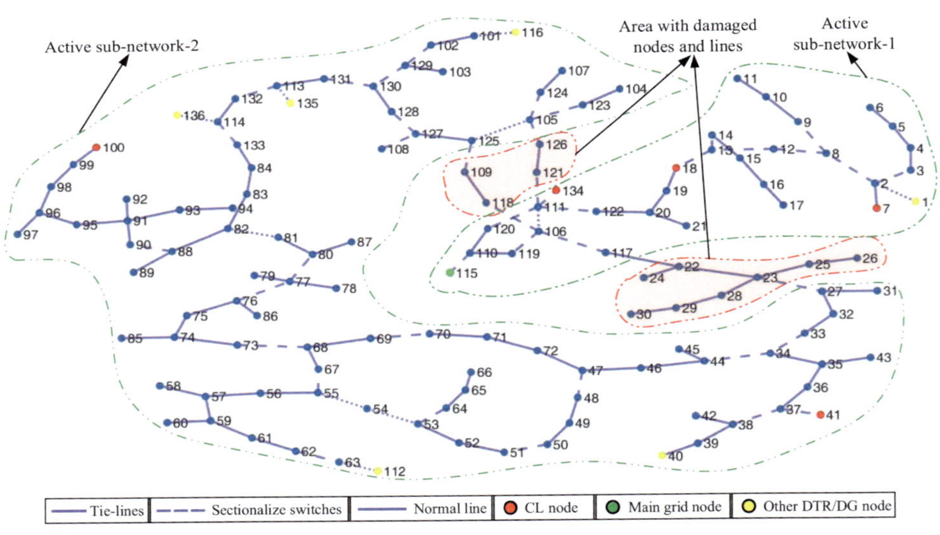

Fig. 9 Graph plot of the Nayabad (N) O/T network for MM placement. Adapted and reproduced with permission from [2]

the load flow convergence with MM as a source and hence the MM is allowed to form the island and feed the CLs when such a circumstance arises. The network operator can enable the switching configuration corresponding to the most resilient FN option, i.e., FN2 having 63.99% resiliency and with MM connected in grid-following mode to the FN for supplying the CLs in the most resilient manner with the available active infrastructure in the sub-network.

Table 12 FNs in the ascending order of composite resiliency with the metric values for active sub-network-1

FN details		Topological					Physical		
FNs	Source node	BCE	OB	SO	PoA &PF	ACPD	LSI	PSI	Composite resiliency (%)
FN 4	137	0	0	0	1	0	1	1	60.51
FN 1	115	0	0	0.250	0.928	0	1	1	60.70
FN 3	1	0.346	0.563	0	1	0.379	0.628	0.864	61.47
FN 5	137	0.115	0.188	0.125	1	0.139	0.923	0.976	63.48
FN 2	115	0.115	0.188	0.375	0.928	0.139	0.923	0.976	63.99

Fig. 10 FN plots

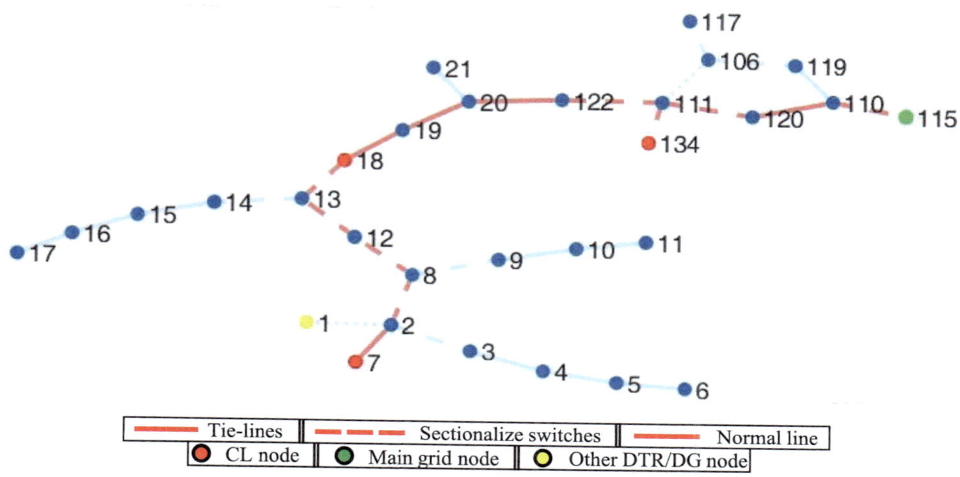

(a) FN with source at 115

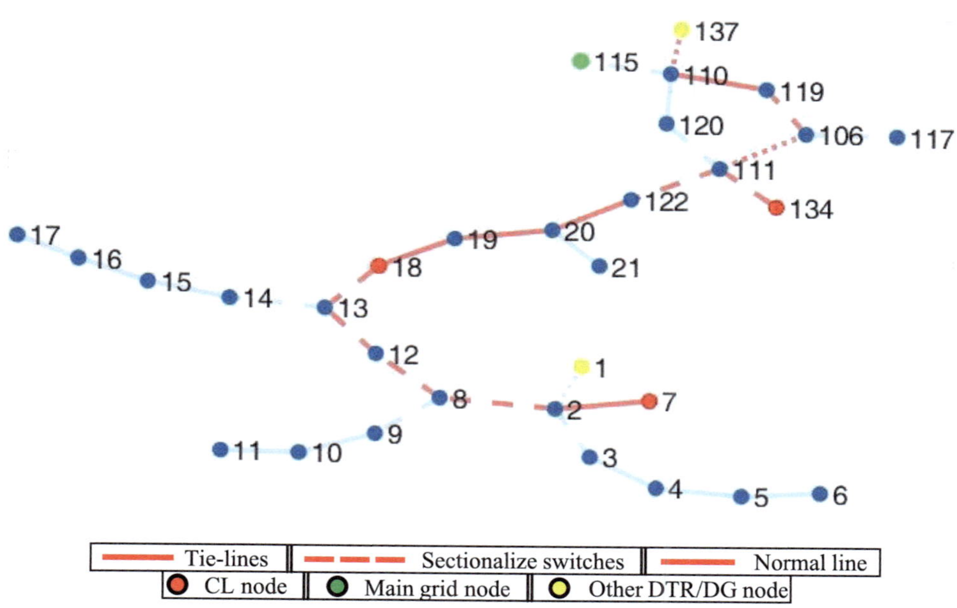

(b) FN with source at 137

5 Conclusion

A framework that performs the assessment, quantification, enabling, and enhancement of resiliency of an OPDN is discussed in this work. When an HILP event occurs, the network operator can respond to it more resiliently by using the resiliency framework to help them make the right decisions. Line strength- and pole strength-related network parameters are among the special set of physical system metrics that are considered in this work. Accurately representing the relative importance and interactions of the various resilience measures is made possible by the resiliency quantification technique. Active sub-networks are dynamically identified depending on the input data, and the suggested methods for operational contingency can manage multi-node damage scenarios during an HILP event. The location and control modalities of MM are determined by the operational resiliency enhancement algorithm. The suggested resiliency structure incorporates MM siting and control, offering a complete solution for upcoming OPDNs with MMs. As long as the necessary physical system metric data for network lines and poles is available, the suggested framework is scalable and applicable to bigger networks.

References

1. Bajpai P, Chanda S, Srivastava AK (2018) A novel metric to quantify and enable resilient distribution system using graph theory and choquet integral. IEEE Trans Smart Grid 9(4):2918–2929
2. Naidu BR, Saini KK, Bajpai P, Chakraborty C (2023) A novel framework for resilient overhead power distribution networks. Int J Electr Power Energy Syst 147:108839
3. Tarjan R (1971) Depth-first search and linear graph algorithms. In: 12th annual symposium on switching and automata theory (swat 1971), pp 114–121
4. Saaty T (2008) Decision making with the analytic hierarchy process. Int J Serv Sci 1:83–98
5. Wu J-Z, Zhang Q (2010) 2-order additive fuzzy measure identification method based on diamond pairwise comparison and maximum entropy principle. Fuzzy Optim Decis Making 9:435–453
6. Grabisch M (1997) k-order additive discrete fuzzy measures and their representation. Fuzzy Sets Syst 92(2):167–189. Fuzzy Measures and Integrals
7. Naidu BR, Bajpai P, Chakraborty C, Malakondaiah M, Kumar BK (2023) Adaptive dynamic voltage support scheme for fault ride-through operation of a microgrid. IEEE Trans Sustain Energy 14(2):974–986

Solar Thermal Heating-based Ammonia Cracking System for Hydrogen Regeneration: Status-Quo and Thermodynamic Analyses

Shubham Jaiswal, Deepank Arya, Ashish Thakur, Kuldeep Awasthi, Sandeep Goli, and Laltu Chandra

Nomenclature

C_R	Heat capacity ratio
$\dot{m}$	Mass flow rate
P	Static pressure
T	Temperature

Greek Symbols

ε	Effectiveness
η	Efficiency
ψ	Waste heat recovery fraction

Abbreviation

CST	Concentrated Solar Thermal
HE	Heat Exchanger
HCRS	High Concentration Radiation Source
HR	Heat Rate
MCRT	Monte Carlo Ray Tracing
NTU	Number of Transfer Units
STS	Solar Tower Simulator
WHR	Waste Heat Recovery

Subscript

c	Conversion/cold
ce	Cold fluid outlet
ci	Cold fluid inlet
ex	Exergy
h	Hot
he	Hot fluid outlet
hi	Hot fluid inlet
max	Maximum
min	Minimum
opt	Optical
rec	Receiver
sh	Solar to hydrogen
th	Thermal

1 Introduction

Alternate energy sources have been promoted to mitigate the use of fossil fuels and the resulting greenhouse gas emissions [1]. For instance, hydrogen with a gravimetric energy density of 120–140 MJ/kg is a promising energy carrier for transportation and manufacturing industries [2]. However, a low volumetric energy density, the high-pressure storage requirements, and production difficulties limit its use. Ammonia (NH_3), with 17.6 wt% of hydrogen, has been recognized as a hydrogen carrier owing to the ease of liquefaction, say at 10 bar and 20 °C. The thermo-catalytic cracking of green ammonia, using renewable energy sources, has been foreseen for onsite hydrogen [3].

S. Jaiswal · D. Arya · A. Thakur · K. Awasthi · S. Goli · L. Chandra (✉)
Department of Sustainable Energy Engineering & Chandrakanta Kesavan Centre for Energy Policy and Climate Solutions, Indian Institute of Technology, Kanpur, Uttar Pradesh, India
e-mail: lchandra@iitk.ac.in

S. Narayanan et al. (eds.), *Materials, Devices and Systems for Sustainability*, IITK Directions 8,
https://doi.org/10.1007/978-981-96-7295-0_4

Chiuta et al. [4] performed experimental and numerical studies to evaluate the performance of a microchannel reactor for hydrogen production through ammonia cracking using a Ru/Al_2O_3 catalyst. They reported a ~99.8% NH_3 conversion rate at 500 Nml/min and 600 °C. Abashar et al. [5] demonstrated that nearly complete ammonia cracking (99.999%) is achievable up to 500 °C using a Ni/Al_2O_3 catalyst in a bench-scale membrane reactor, with a reaction pressure of 3.65 MPa and a sweep gas pressure of 0.10 MPa. Thermo-catalytic ammonia cracking requires high temperatures, which can be efficiently supplied by concentrated solar thermal (CST) systems, such as parabolic troughs and heliostats. A few numerical investigations have been reported with parabolic trough collectors. However, the use of thermally efficient, heliostat-based, solar thermal systems has not been explored. Recently, the concept of a novel solar tower-based simulator has been developed for hydrogen regeneration by thermo-catalytic ammonia cracking [6]. The need for detailed analyses, including waste heat recovery and reuse sub-system, is underlined for this system. Thus, the following aspects are presented in this chapter:

(a) a review of the current status of the solar thermal-based ammonia cracking systems,
(b) the thermodynamics-based assessment of the solar tower simulator system,
(c) a waste heat recovery and reuse sub-system for reutilization of heat, and
(d) the development of a secondary reflector for enhancing the overall system efficiency.

2 Status-Quo of Solar Thermal-Based Ammonia Cracking Systems

CST systems are extensively researched for high-temperature thermo-catalytic ammonia cracking. Line-focusing CST technology provides up to 100 Suns and 400 °C, while point-focusing offers up to 1500 Suns and over 700 °C, ideal for high conversion rates [7]. Wang et al. [8] designed a line-focusing reactor using a nickel-based catalyst within a parabolic trough. Carden et al. [9] developed a thermochemical energy storage-based CST system. Lovegrove et al. [10] created a 2D model of a homogeneous fixed catalytic bed reactor to analyze the exergy efficiency, and validated the same with experimental data. Meanwhile, Williams [11] designed ammonia dissociation reactors using paraboloid dish solar thermal systems. Hu and Wang [12] created a 1D thermodynamic model for a CST-driven tubular reactor. Wang et al. [13] proposed a solar ammonia cracking system, including a hydrogen permeation membrane reactor, to optimize thermal performance and environmental impact. Additionally, numerous studies have emphasized the use of waste heat recovery and reuse sub-systems to improve efficiency and reduce emissions [9, 10, 13]. Devkota et al. [14] explored hydrogen production from ammonia cracking using a multi-catalytic packed bed reactor with waste heat utilization. Seklani and Bicer [15] proposed a zero-emission system producing hydrogen, ammonia, and urea from natural gas, integrating waste heat recovery in a Rankine cycle for enhanced energy efficiency.

3 Solar Tower Simulator for Ammonia Cracking

3.1 System Design

Figure 1a depicts a heliostat-based solar tower system for thermo-catalytic ammonia cracking. In this setup, ammonia is supplied to a receiver exposed to concentrated solar irradiance, leading to hydrogen regeneration through thermo-catalytic cracking. To replicate this process at a laboratory scale, an artificial radiation-based solar tower simulator (STS) has been developed, as shown in Fig. 1b.

The STS includes (a) a flow-loop for ammonia heating and subsequent catalytic cracking to generate hydrogen, (b) a high-concentration, solar-like, radiation source (HCRS), and (c) an air-cooling system for the HCRS and its internal components. Figure 1b illustrates various parts of the ammonia heating loop, including a preheater, which is designed to utilize the available waste heat. The HCRS installation allows for different heat flux levels, based on varying ammonia mass flow rates, in the heating loop. For a thermodynamic assessment, the maximum ammonia mass flow rate is set to 1 kg/h, and net solar-to-hydrogen (η_{sh}), first-law (η_{th}), and exergy (η_{ex}) efficiency, including waste heat recovery fraction, have been evaluated using the adopted expressions from Wang et al. [8]. The calculations have been presented in Sect. 4.

3.2 Waste Heat Recovery and reuse Sub-system

A waste heat recovery (WHR) and reuse sub-system has been developed for preheating the subcooled liquid ammonia to the saturated ammonia vapor. The designed WHR and reuse sub-system comprises two shell-and-tube counterflow heat exchangers (HE_1 and HE_2) (see Fig. 2). The WHR and reuse sub-system enables heat transfer from hot ammonia at 600 °C to the preheater. Therminol-55 is used as a heat transfer fluid from HE_1 to HE_2. A validated mathematical model has been developed and implemented using MATLAB for this sub-system. The analysis assumes (i) a steady-state condition, (ii) an adiabatic system, and (iii) no internal heat generation.

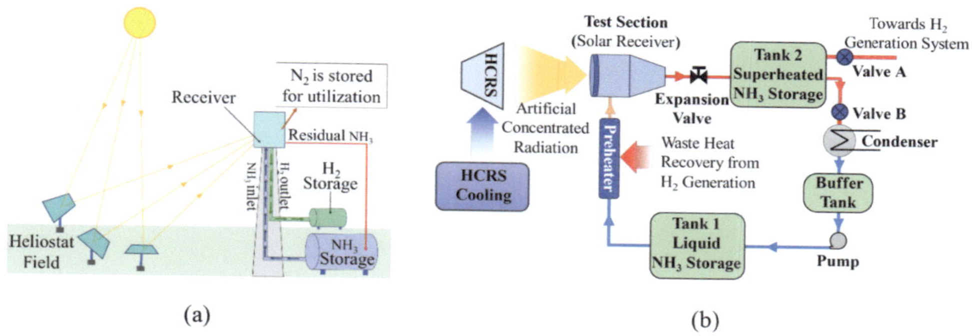

Fig. 1 Schematic of **(a)** a volumetric receiver-based solar tower for ammonia cracking [6], and **(b)** solar tower simulator for heating of ammonia ($P_{max} = 10$ bar) [16]

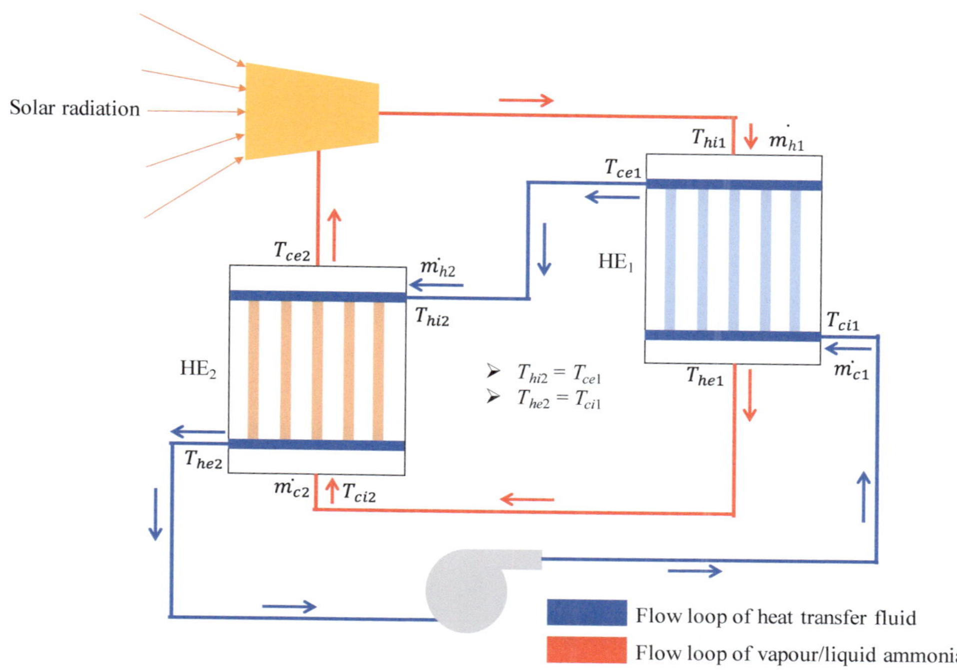

Fig. 2 A schematic of the WHR and reuse sub-system

3.3 Design of a Secondary Collector

The investigation assumes that the HCRS system focus onto a circular spot with a 20 mm diameter, smaller than the receiver aperture of 30 mm [6]. The HCRS comprises seven 350 We Xe-short-arc lamp and reflector arrangment. The estimated incident power onto a 30 mm circular spot is 981 W, and thus an optical efficiency of ~40% can be achieved. The use of a secondary reflector is expected to enhance the collection and the corresponding optical efficiency [13]. Figure 3 illustrates such a secondary reflector–receiver arrangement. The reflector has a circular (secondary) aperture of 70 mm diameter. A suitable location for the secondary aperture is crucial to reduce the spillage, which has been investigated using the Monte Carlo-based ray tracing simulator in ANSYS. The findings have been discussed subsequently.

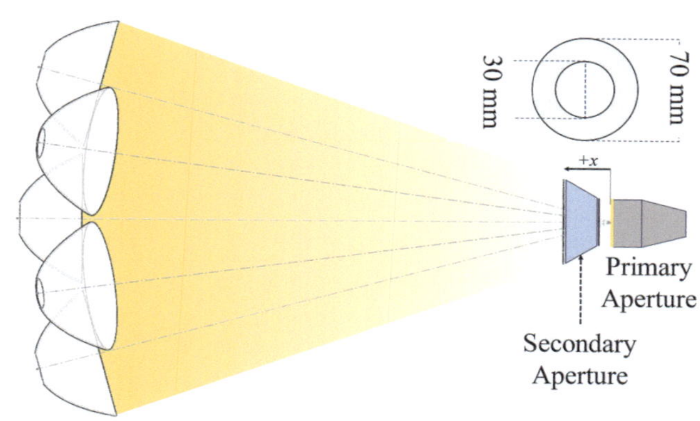

Fig. 3 Schematic of the secondary collector receiver arrangement

4 Results and Discussions

The net solar-to-hydrogen generation efficiency (η_{sh}) depends on optical efficiency (η_{opt}), waste heat recovery fraction (ψ), receiver efficiency (η_{rec}), conversion efficiency (η_c),

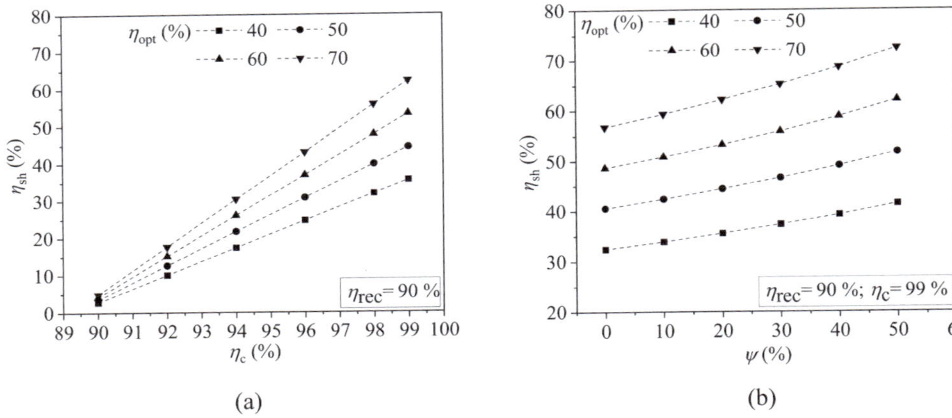

Fig. 4 Variation of the net solar-to-hydrogen efficiency (η_{sh}) with **(a)** conversion efficiency (η_c) for $\eta_{rec} = 90\%$, and **(b)** waste heat recovery fraction (ψ) for $\eta_c = 99\%$ and $\eta_{rec} = 90\%$

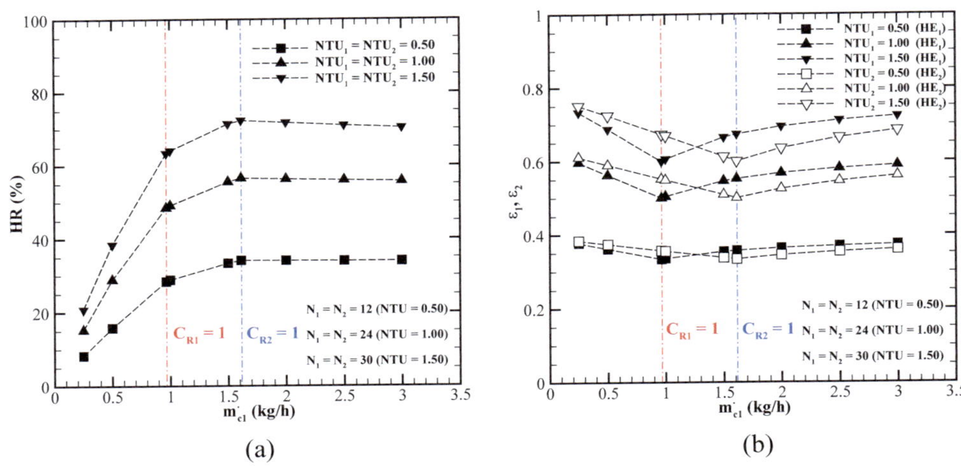

Fig. 5 The calculated **a** HR as a function of $\dot{m}_{c1}$ and NTU; **b** effectiveness of HE_1 (ε_1) and HE_2 (ε_2) as a function of $\dot{m}_{c1}$ and NTU

and minimum operating pressure (P_{min}). Figure 4a reveals the calculated η_{sh} for $\eta_{rec} = 90\%$, $40 \leq \eta_{opt} \leq 70\%$, and $90 \leq \eta_c \leq 99\%$ at $P_{min} = 9$ bar. The η_{sh} increases with increasing η_c and η_{opt}. For instance, at $\eta_{opt} = 40\%$, η_{sh} increases from 2.8 to 35.5% for $90 \leq \eta_c \leq 99\%$, and at $\eta_c = 99\%$, η_{sh} increases from 35.5 to 62.1% for $40 \leq \eta_{opt} \leq 70\%$. Figure 4b illustrates the effect of ψ on η_{sh} for $40 \leq \eta_{opt} \leq 70\%$, $\eta_{rec} = 90\%$, and $\eta_c = 99\%$. The η_{sh} increases by 27% as ψ increases from 0 to 50%. These findings highlight the potential of solar tower technology for large-scale hydrogen regeneration by thermo-catalytic ammonia cracking.

Preliminary calculations were conducted to estimate the waste heat recovered (HR) in HE_1 as a fraction of the required amount for vaporizing liquid ammonia in the preheater stage (HE_2). Furthermore, the corresponding effectiveness of HE_1 (ε_1) and HE_2 (ε_2) was determined. Both HE_1 and HE_2 are assumed to have the same number of tubes for simplicity. C_{R1} and C_{R2} are the ratio of minimum-to-maximum heat capacity rates in HE_1 and HE_2, respectively. Liquid ammonia enters the preheater (HE_2) at 15 °C and vaporizes at 25 °C at a pressure of 10 bar [17], and the required latent heat of vaporization is 1167.42 kJ/kg [18]. Thus, for the target ammonia mass flow rate of 1 kg/h, the rate of heat input to the preheater is 338 W.

Figure 5a depicts the variation of HR with coolant mass flow rate ($\dot{m}_{c1}$) and number of heat transfer units (NTU), and Fig. 5b provides the corresponding ε_1 and ε_2 to achieve the HR for a given $\dot{m}_{c1}$ and NTU. Figure 5a shows that the HR increases with increasing $\dot{m}_{c1}$ up to $C_{R2} = 1$, beyond which it remains constant. Higher NTU values further enhance HR due to augmented ε_1 and ε_2, as depicted in Fig. 5b. The maximum HR of 72% occurs at $NTU = 1.50$ and a mass flow rate of 1.61 kg/h, with ε_1 and ε_2 values of 0.67 and 0.60, respectively. Design calculations for $NTU = 1.50$ and $C_{R2} = 1$ suggest 30 tubes per heat exchanger, with tube lengths of 287 mm for HE_1 and 88.5 mm for HE_2, for a tube with an 8 mm inner diameter.

MCRT-based ray tracing analysis indicates that the 30 mm aperture of the solar receiver captures 980 W of the incident power leading to an overall optical efficiency of ~40%. Figure 6 shows the computed values of intercepted power at the secondary and receiver aperture. Following are the major findings including a secondary reflector:

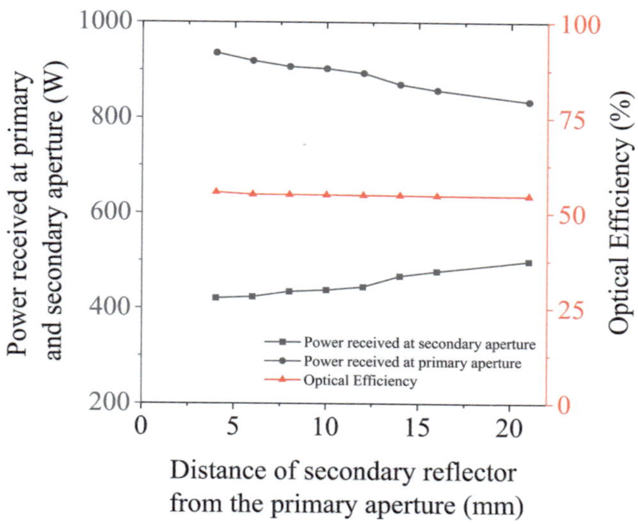

Fig. 6 MCRT-based intercepted power at the primary and secondary aperture

a. The intercepted power by the secondary aperture increases from ~425 to ~500 W with respect to the distance (4–21 mm) from the primary aperture.
b. The total power, intercepted by the secondary reflector and at the receiver aperture, ranges between 1335 and 1370 W. Thus, the optical efficiency increases by up to 15%

5 Conclusions

Ammonia has been recognized as a potential, green, hydrogen carrier. A novel concept of heliostat-based solar tower simulator has been developed for thermo-catalytic ammonia cracking, including a waste heat recovery mechanism. Thermodynamic analyses, ε-NTU-based heat exchanger design calculations, and Monte Carlo ray tracing simulations have been performed for the design of the simulator. The major findings are as follows:

a. Solar-to-hydrogen efficiency (η_{sh}) can reach nearly 62% when the optical efficiency (η_{opt}) is 70% and the conversion efficiency (η_c) is 99%. Additionally, incorporating waste heat recovery can increase η_{sh} by 27% when the waste heat recovery fraction (ψ%) reaches up to 50%. These results demonstrate that optimizing optical efficiency, conversion efficiency, and waste heat recovery are crucial for the system.
b. Preliminary design calculations for the waste heat recovery (WHR) and reuse sub-system indicate that the maximum heat rate recovered (HR) is 72% at an *NTU* of 1.50 and a coolant mass flow rate of 1.61 kg/h. The corresponding effectiveness values for HE_1 and HE_2 are 0.67 and 0.60, respectively.
c. The use of a secondary reflector in the HCRS enhances optical efficiency by about 15%. This will enhance the overall system efficiency.

Acknowledgements The authors gratefully acknowledge the support received from IIT Kanpur, Chandrakanta Kesavan Center for Energy Policy and Climate Solutions vide grant number DORA /DORA/2021136D, and Kotak School of Sustainability vide grant number DORA /DORA/2023508J. The authors also acknowledge the received support from the technical staff members of the laboratory.

References

1. Cha J et al (2021) An efficient process for sustainable and scalable hydrogen production from green ammonia. Renew Sustain Energy Rev 152:111562. https://doi.org/10.1016/j.rser.2021.111562
2. Wang B, Li T, Gong F, Othman MHD, Xiao R (2022) Ammonia as a green energy carrier: electrochemical synthesis and direct ammonia fuel cell—a comprehensive review. Fuel Process Technol 235:107380. https://doi.org/10.1016/j.fuproc.2022.107380
3. MacFarlane DR et al (2020) A roadmap to the ammonia economy. Joule 4(6):1186–1205. https://doi.org/10.1016/j.joule.2020.04.004
4. Chiuta S, Everson RC, Neomagus HWJP, Bessarabov DG (2016) Hydrogen production from ammonia decomposition over a commercial Ru/Al2O3 catalyst in a microchannel reactor: experimental validation and CFD simulation. Int J Hydrogen Energy 41(6):3774–3785. https://doi.org/10.1016/j.ijhydene.2015.12.130
5. Abashar MEE, Al-Sughair YS, Al-Mutaz IS (2002) Investigation of low temperature decomposition of ammonia using spatially patterned catalytic membrane reactors. Appl Catal A Gen 236(1–2):35–53. https://doi.org/10.1016/S0926-860X(02)00272-7
6. Arya D, Awasthi K, Hedau G, Chandra L (2024) A novel solar tower simulator for hydrogen regeneration by thermo-catalytic cracking of ammonia: energy, exergy, and CFD investigations. Appl Therm Eng 256:124141. https://doi.org/10.1016/j.applthermaleng.2024.124141
7. Zhang HL, Baeyens J, Degrève J, Cacères G (2013) Concentrated solar power plants: review and design methodology. Renew Sustain Energy Rev 22:466–481. https://doi.org/10.1016/j.rser.2013.01.032
8. Wang B, Kong H, Wang H, Wang Y, Hu X (2019) Kinetic and thermodynamic analyses of Mid/Low-temperature ammonia decomposition in solar-driven hydrogen permeation membrane reactor. Int J Hydrogen Energy 44(49):26874–26887. https://doi.org/10.1016/j.ijhydene.2019.08.175
9. Carden PO, Williams OM (1978) The efficiencies of thermochemical energy transfer. Int J Energy Res 2(4):389–406. https://doi.org/10.1002/er.4440020406
10. Lovegrove K, Luzzi A, Mccann M, Freitag O (1999) Exergy analysis of ammonia-based solar thermochemical power systems. Sol Energy 66(2):103–115. https://doi.org/10.1016/S0038-092X(98)00132-7
11. Williams OM (1980) Design and cost analysis for an ammonia-based solar thermochemical cavity absorber. Sol Energy 24(3):255–263. https://doi.org/10.1016/0038-092X(80)90482-X
12. Hu T, Wang Y (2020) Effect of operating and geometrical parameters on ammonia decomposition in a tubular reactor driven by concentrating solar power. J Energy Eng 146(4). https://doi.org/10.1061/(ASCE)EY.1943-7897.0000664
13. Wang R, Sun J, Hong H, Liu Q (2017) An on-site test method for thermal and optical performances of parabolic-trough loop for

utility-scale concentrating solar power plant. Sol Energy 153:142–152. https://doi.org/10.1016/j.solener.2017.05.053
14. Devkota S et al (2023) Process design and optimization of onsite hydrogen production from ammonia: reactor design, energy saving and NOX control. Fuel 342:127879. https://doi.org/10.1016/j.fuel.2023.127879
15. Seklani Y, Bicer Y (2024) Comprehensive assessment of solar cracking of natural gas for the production and transportation of turquoise energy carriers. Int J Hydrogen Energy 51:536–558. https://doi.org/10.1016/j.ijhydene.2023.07.284
16. Arya D, Awasthi K, Hedau G, Chandra L (2021) Concentrated solar thermal-based hydrogen generation: some recent findings and a proposal for experiment setup, Chapter 16, pp 205–215. https://doi.org/10.1007/978-981-99-9009-2_16
17. National Institute of Standards and Technology (NIST), NIST Standard Reference Database Number 69, https://webbook.nist.gov/cgi/cbook.cgi?Name=Ammonia&Units=SI
18. Circular of the bureau of standards—tables of thermodynamic properties of ammonia (142), 1st ed., vol. 142. Washington, D.C.: Washington Government Printing Office, 1923

S-scheme Heterojunctions: An Innovative Design of a Powerful Photocatalyst

Abhishek Gupta, Rahul Gupta, and Nishith Verma

1 Introduction

Photocatalysis, driven by the interaction of light with semiconductor materials, has garnered significant attention because of its potential for power generation and environmental remediation. Traditional heterojunctions, classified as type-II and Z-scheme, involve the integration of two semiconductors with staggered band positions [1]. However, such configurations often face limitations in efficiently utilizing solar energy for photocatalytic reactions. Type-II heterojunctions exhibit inefficient charge transfer, whereas Z-scheme heterojunctions suffer from inadequate spatial segregation of photogenerated holes and electrons, ultimately restricting their efficiency [2]. To address these shortcomings, the focus has recently shifted toward S-scheme heterojunctions. The S-scheme configurations involve the integration of two semiconductors with aligned or partially overlapped energy levels. This arrangement promotes enhanced charge transfer and separation, contributing to improved redox reactions in the light irradiation [3].

Two different types of photocatalysts are used in an S-scheme heterojunction. One has a relatively more positive valence band (VB), known as oxidizing photocatalyst (OP), and the other has a relatively more negative conduction band (CB), known as reducing photocatalyst (RP). RP has relatively more negative CB and VB potentials, along with a lower Fermi level, than OP. When RP and OP come in contact with each other, e^- from RP spontaneously flow toward OP because of the difference in their Fermi levels. The charge migration leads to the formation of an e^--deficient region in RP and an e^--rich region in OP near the interface. The redistribution of charges at the interface establishes an internal electric field directed from RP to OP, facilitating efficient charge segregation and enhancing photocatalytic performance [4]. The electric field further promotes the transfer of photogenerated e^- from OP to RP [5]. Additionally, when OP and RP are in contact, their Fermi energies align, causing an upward shift in the Fermi level of OP and a downward shift in that of RP [6]. The alignment leads to band bending, enabling the recombination of photoinduced e^- in the CB of OP with holes in the VB of RP within the interfacial region [7]. This recombination is primarily driven by Coulombic attraction. The combined effects of the internal electric field, band bending, and Coulombic forces work synergistically to facilitate the interaction and recombination of e^- from the CB of OP and h^+ from the VB of RP. The recombination process effectively eliminates excess e^- and h^+, while preserving the highly energetic photogenerated e^- in the CB of RP and h^+ in the VB of OP. [8]. The retained e^- and h^+ are then available to participate in the photocatalytic reactions such as water splitting, nitrogen and nitrates reduction to produce H_2 and NH_3, respectively [9–11].

In recent studies, novel S-scheme heterojunctions have been synthesized by using graphitic carbon nitride ($g\text{-}C_3N_4$/gCN) as RP and $NiSnO_3$ (NSO), Fe_2SnO_4 (FSO), and $NiTiO_3$ (NT) as OP [12–15]. The composite heterojunctions are synthesized using the facile ultra-sonification, followed by calcination and solvothermal techniques. The synthesized heterojunctions show an excellent photocatalytic activity in hydrogen production and nitrogen fixation.

2 $NiSnO_3$/$g\text{-}C_3N_4$ for H_2 Production

Widespread reliance on conventional energy sources has raised grave environmental concerns, particularly global warming. Hydrogen, emerging as a viable alternative to traditional fuels such as gasoline and coal, is pivotal in

A. Gupta · R. Gupta · N. Verma (✉)
Department of Chemical Engineering, Department of Sustainable Energy Engineering, and Centre for Environmental Science and Engineering, Indian Institute of Technology Kanpur, Kanpur, India
e-mail: nishith@iitk.ac.in

S. Narayanan et al. (eds.), *Materials, Devices and Systems for Sustainability*, IITK Directions 8,
https://doi.org/10.1007/978-981-96-7295-0_5

mitigating these issues. However, current hydrogen production methods face challenges related to the environmental sustainability and economic viability. The photocatalytic hydrogen production via water splitting reaction is an efficient route for harnessing the inexhaustible solar energy. Recently, Pandey et al. [12] have synthesized an innovative $NiSnO_3/g\text{-}C_3N_4$ (NSO-gCN) S-scheme, where NSO nanoparticles (NPs) are intricately anchored onto gCN sheets, creating an effective interface coupling for the S-scheme heterojunction. UV–vis diffuse reflectance spectra (UV-vis DRS) analysis of NSO-gCN heterojunction revealed an absorption edge at 463 nm, indicating its exceptional capability to absorb visible light. X-ray photoelectron spectra (XPS) revealed shifts in the binding energies of the constituent elements, corroborating the band bending of the photocatalysts and confirming the S-scheme-based heterojunction between NSO and gCN. Electrochemical impedance spectra (EIS) of NSO-gCN validated the efficacy of the S-scheme charge transfer mechanism, with the least arc radius indicating a minimal charge transfer resistance in the composite. Figure 1 is the graphical representation of the photoinduced hydrogen generation by NSO-gCN photocatalyst, where triethanolamine (TEOA) is used as the sacrificial agent for scavenging holes to prevent their recombination with the photogenerated electrons. When used for the photocatalytic water splitting reaction, NSO-gCN demonstrated a significant H_2 generation rate of 646 μmol g^{-1} h^{-1}, achieving 2.1 times higher performance compared to gCN alone. Impressively, this enhanced performance was sustained over five additional cycles, emphasizing the stability of the S-scheme NSO-gCN heterojunction.

3 Ruddlesden–Popper Phase Fe_2SnO_4-g-C_3N_4 for H_2 Production

While a large surface area is necessary for enhancing light harvesting and catalytic activity in photochemical reactions, traditional 3D perovskites face limitations because of their enclosed cage-like structures. The 2D Ruddlesden–Popper phase perovskites with their layered architecture can be a potentially alternative material to be studied. Their exposed, sheet-like surfaces offer significantly large accessible areas for reactants and light-harvesting molecules to interact with an enhanced light absorption. Also, the modularity of the 2D structure allows for precise tuning of the materials electronic properties, enabling an optimization of specific target reactions. Each layer in an Ruddlesden–Popper phase perovskite acts as a potential active site for water splitting reaction. This abundance compared to 3D perovskites translates to a higher number of reaction initiation points, boosting overall hydrogen production efficiency. Think of it as having multiple factories operating simultaneously instead of just one. The combination of high surface area, tunable electronic properties, and abundant active sites renders the 2D Ruddlesden–Popper phase perovskites an ideal candidate for S-scheme heterojunction photocatalysts, outperforming their 3D counterparts for several photochemical applications. Gupta et al. [13] fabricated an S-scheme heterojunction by integrating the A_2BX_4-type Ruddlesden–Popper perovskite Fe_2SnO_4 (FSO) with gCN (Fig. 2). The binary composite was synthesized by ultrasonicating an aqueous mixture of FSO and gCN, followed by calcination, the resulting material referred to as FCN. The synthesized materials were characterized using various techniques to analyze

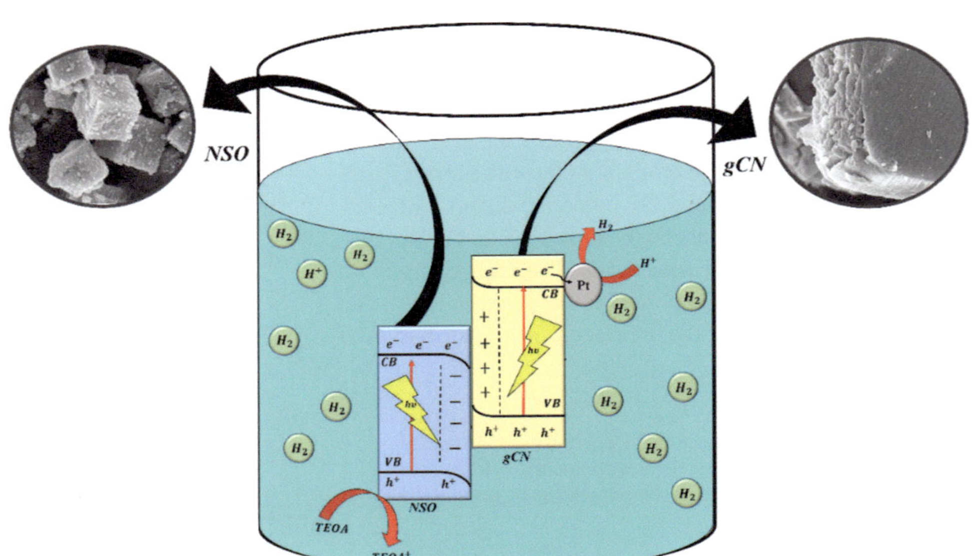

Fig. 1 Graphical representation of photocatalytic H_2 production by the $NiSnO_3/g\text{-}C_3N_4$ S-scheme heterojunction. Reprinted from Ref. [12], Copyright (2023), with permission from Elsevier

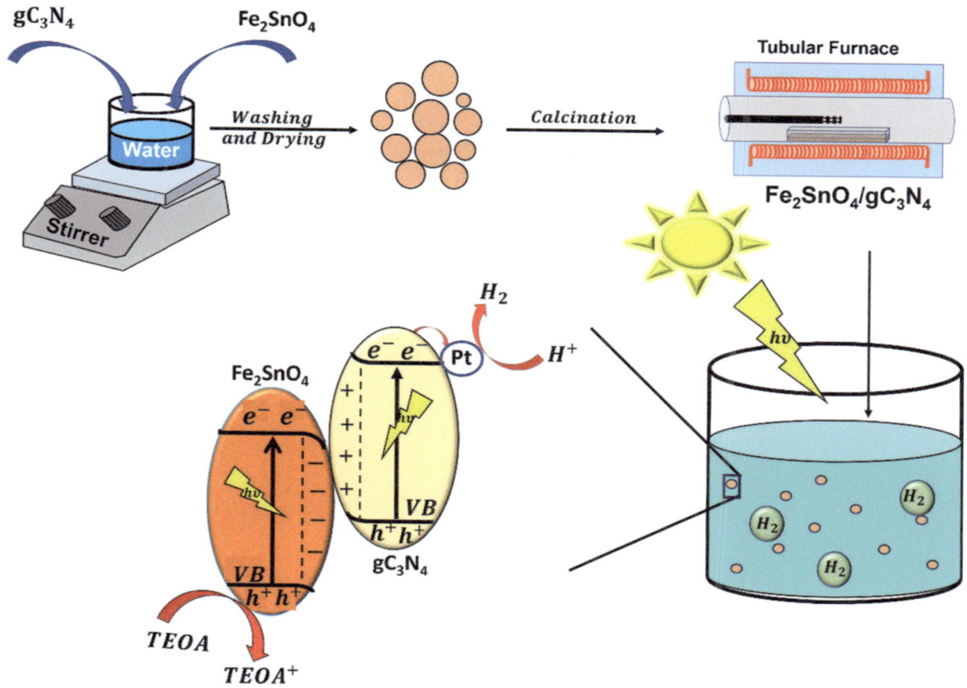

Fig. 2 Graphical representation of photocatalytic H_2 production by $Fe_2SnO_4/g\text{-}C_3N_4$ S-scheme heterojunction. Reprinted from Ref. [13], Copyright (2024), with permission from Elsevier

the physicochemical, optical, and electrochemical properties, confirming the establishment of an S-scheme mediated charge transfer between FSO and gCN. The synthesized materials (FSO, gCN, and FCN) were tested for photocatalytic H_2 production via water splitting reaction in the visible light. FCN heterojunction demonstrated a significant H_2 production rate of 1204 μmol g^{-1} h^{-1} surpassing that of pristine FSO and gCN by factors of 133 and 3.7, respectively. Such improved activity was ascribed to the high specific surface area and improved charge transfer facilitated by the 2-D Ruddlesden–Popper phase-based S-scheme heterojunction.

4 Anthraquinone (AQ)-Integrated $NiTiO_3$/$g\text{-}C_3N_4$ for H_2 Production

Enhancing S-scheme photocatalytic system performance involves incorporating promoters with suitable redox potentials, which can boost light harvesting and charge separation at the heterojunction surface. A highly effective method for accelerating reaction kinetics and structuring nanoenvironments is the anchoring of semiconductors on organic redox molecules. AQ, known for its e^- trapping abilities, stands out as an effective promoter because of its unique redox properties and relatively higher conductivity. AQ facilitates electron transfer by linking functional groups with semiconductor materials. Alam et al. [14] synthesized an AQ-integrated $NiTiO_3/g\text{-}C_3N_4$ (AQ-NT-gCN) S-scheme heterojunction. AQ-NT-gCN was synthesized using a solvothermal method. The S-scheme-mediated charge transfer mechanism was validated through various radical trapping experiments. AQ played a pivotal role in improving charge segregation efficiency, as demonstrated by the photoluminescence (PL) analysis. Figure 3 shows the schematic representation of the photocatalytic H_2 production by the synthesized heterojunction. The ternary S-scheme heterojunction exhibited a significant H_2 production rate of 576 μmol g^{-1} h^{-1}, outperforming the binary composite NT-gCN with a H_2 production rate of 470 μmol g^{-1} h^{-1}, indicating a significant 23% increase in the overall activity.

5 Photofixation of N_2 to NH_3 Over $NiSnO_3/g\text{-}C_3N_4$

NH_3 is a vital precursor for nitrogen-rich synthetic fertilizers, with an annual global production surpassing 180 million metric tons. The dominant industrial method, viz., the Haber–Bosch process, operates at high temperatures ranging from 300 to 400 °C and pressures between 150 and 250 bar. The process contributes to 2% of worldwide energy consumption and emits 500 tons of CO_2 each year. Consequently, developing an eco-friendlier method for NH_3 production has become imperative. Photocatalysis harnesses energy from the sun to facilitate the generation of e^- and h^+

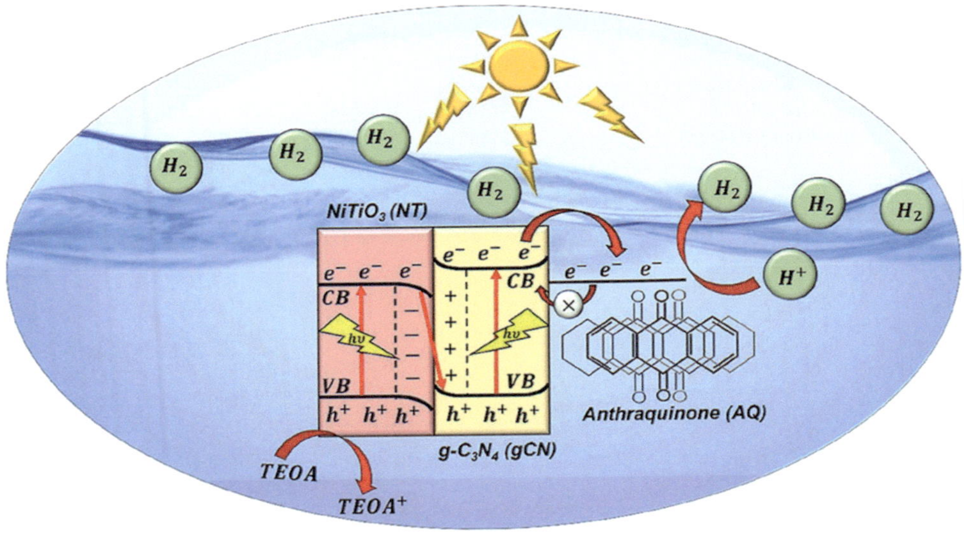

Fig. 3 Schematic representation of the photoinduced H_2 generation of the S-scheme-based AQ-NT-gCN heterojunction. Reprinted from Ref. [14], Copyright (2023), with permission from Elsevier

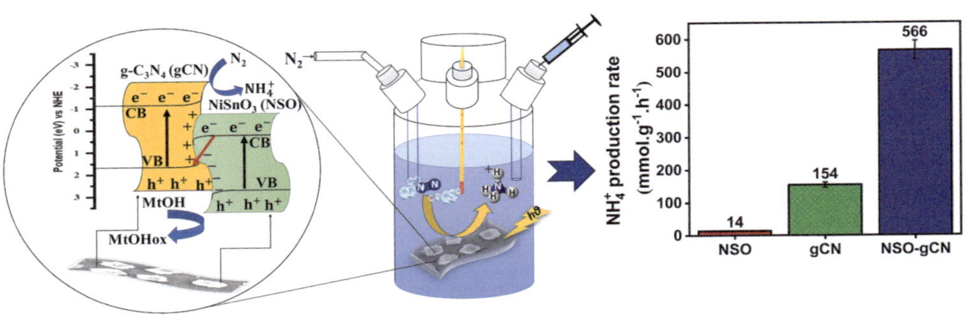

Fig. 4 Schematic representation of the photofixation of N_2 to ammonia by the S-scheme NSO-gCN heterojunction. Reprinted from Ref. [15], Copyright (2024), with permission from Elsevier

in the CB and VB of semiconductor materials, respectively. The electrons activate inert nitrogen molecules, offering a promising, energy-efficient, and environmentally friendly alternative for nitrogen fixation to ammonia. The recently developed S-scheme heterojunction materials have garnered significant attention for their capability to facilitate the faster charge segregation and transport. This is attributed to the electric field generated at the heterojunction region. The S-scheme-mediated charge transfer mechanism-based heterojunction enhances the generation of highly energetic photogenerated e^- and h^+, with the photogenerated h^+ retained in the VB of the OP and e^- in the CB of the RP. Gupta et al. [15] successfully synthesized an S-scheme heterojunction by combining the cost-effective NSO perovskite with gCN. NSO-gCN heterojunction demonstrated exceptional nitrogen fixation activity under visible light, attaining an outstanding NH_3 generation rate of 566 μmol g^{-1} h^{-1}. Additionally, NSO-gCN demonstrated remarkable stability, with a minor reduction in activity observed after three reaction regeneration cycles of the photocatalytic reaction. The strong photogenerated electrons efficiently reduce N_2 molecules to ammonia (Fig. 4).

6 Materials Synthesis

gCN is synthesized by the melamine polymerization at 550 °C in N_2 in an electrical furnace. A two-step synthesis is implemented for synthesizing NSO NPs. The aqueous solutions of nickel nitrate and sodium stannate are prepared and mixed. 1 mL of ammonia is added in the mixture for regulating the pH of the mixture. The mixture is then subjected to a 24 h hydrothermal reaction at 180 °C, enabling the controlled crystallization of nickel stannate oxide NPs. Subsequent washing and calcination steps further purified and stabilized the material. FSO was synthesized by preparing an equimolar mixture of iron nitrate and sodium stannate in separate solutions (A and B). Solution B was slowly mixed to solution A with constant stirring, and NH_3 was then added to adjust the pH. The resultant mixture is stirred for next 30 min and subjected to hydrothermal treatment in the Teflon-lined vessel at 180 °C for 12 h. The material is washed three times with H_2O and C_2H_5OH, then air-dried at 30 °C for 8 h. The dried material is then heated for 4 h in a tubular furnace at 450 °C. The resulting material is powdered using mortar and pestle and termed as FSO. NT NPs are produced by the solvothermal technique using ethylene glycol as the solvent. Equimolar quantities of nickel nitrate and titanium isopropoxide are dissolved separately in ethylene glycol. The mixture is stirred for 2 h to promote precursor complexation, then undergoes hydrothermal treatment in the Teflon-lined vessel at 180 °C for 24 h. This controlled reaction environment yields well-defined NT nanostructures. NSO-gCN and FCN binary composites are synthesized by dispersing uncalcined NSO and FSO with gCN at fixed proportions in water. The mixture is homogenized through vigorous stirring and sonication, then dried and heated at 450 °C for 3 h. This facilitates strong interfacial contact between the components, crucial for efficient charge carrier transfer within the S-scheme heterojunction.

7 Materials' Physicochemical and Optical Characterization

The synthesized photocatalysts were analyzed by physicochemical, optical, and electrochemical characterization techniques. The crystalline planes in the photocatalysts were determined using X-ray diffraction (XRD) analysis which was performed using an X'pert3 Powder PANalytical instrument (USA). The diffraction spectra were recorded over a 2θ range of 10 to 100° with a scanning speed of 3° per min. Raman spectra were acquired using the Alpha 300R Witec (Germany) instrument in the range 100–1400 cm^{-1} with 532 nm excitation to confirm the heterojunction formation. Fourier-transform infrared (FTIR) spectroscopic analysis was performed using the Bruker Tensor 27 spectrophotometer (Germany) to determine functional groups in the materials. The specific surface area of the materials was determined by analyzing the Brunauer–Emmett–Teller (BET, AS1-C Quantachrome instrument, USA) curve. The Carry UMS 7000 Agilent instrument (USA) was used to record the UV–vis DRS of the samples, with $BaSO_4$ as the reference powder. The PL spectra were recorded using the Agilent Carry Eclipse instrument (USA) with 330 nm as the excitation wavelength. XPS measurements of the samples were taken using the K-th PHI 5000 Versa Prob II instrument (USA). EIS analysis and transient photocurrent measurements were performed to assess charge transfer resistance of the photocatalysts. The electrochemical characterization was performed in the electrochemical workstation (AUTOLAB PGSTAT-302 N, Netherlands). Ag/AgCl and Pt mesh were used as reference- and counter-electrode, respectively. The working electrode was prepared by coating the slurry of photocatalytic material over indium tin oxide electrode. The slurry of photocatalyst was prepared by mixing 10 mg material and 2 mg binding material (polyvinylidene fluoride) in 10 mL solvent (IPA). The slurry was drop-cast over the conductive side of indium tin oxide electrode having the dimensions of 1 X 1 cm^2. The analysis was performed under the illumination of visible light in a 0.1 M Na_2SO_4 solution.

Figure 5a exhibits the XRD spectra of NSO-gCN and its constituents. gCN shows the distinct peaks at 13.1 and 27.3°, corresponding to the (1 0 0) and (0 0 2) crystallographic planes, respectively. The prominent peak at 27.3° shows the interlayer stacking of aromatic segments, which is crucial for promoting charge carrier mobility. NSO shows its characteristic fingerprint with the pronounced peaks at 34.0, 54.36, and 65.34°, signifying its unique crystallographic structure. The harmonious coexistence of these individual signatures within the NSO-gCN composite serves as a compelling evidence for the successful formation of the heterojunction. Raman spectra in Fig. 5b further confirm the successful synthesis of the binary composite. The low-intensity, characteristic bands provide additional validation for the successful integration of NSO and gCN, paving the way for synergistic photocatalytic activity. The N_2 adsorption–desorption isotherm in Fig. 5c indicates the mesoporous characteristics for gCN, NSO, and NSO-gCN, represented by a type IV isotherm. UV-vis DRS data presented in Fig. 5d and e reveal the light absorption range and bandgap energy, respectively. The absorption edges for NSO and gCN were located at the wavelengths of 550 and 455 nm, respectively. However, the absorption edge for the NSO-gCN is found to be located at the wavelength of 463 nm, indicating intermediate light absorption in comparison to its constituents. The bandgaps calculated form Tauc plots are determined to be 2.57 and 2.77 eV for NSO and gCN, respectively. The positions of VB and CB of an individual photocatalyst (viz., NSO and gCN) using the data obtained from Tauc plot analysis and the following empirical equations:

$$E_{CB} = X - E_e - 0.5\ E_g$$

$$E_{VB} = E_{CB} + E_g$$

where E_{CB}, X, E_e, E_{VB}, and E_g are the position of CB, absolute electronegativity, energy of free electrons on the hydrogen scale ($E_e = 4.5$ eV), position of VB, and bandgap of material, respectively. E_{CB} values for NSO and gCN are calculated to be 0.26 and -1.15 eV, respectively, while the E_{VB} values are determined to be 2.79 and 1.62 eV for NSO and gCN, respectively. Charge carrier separation ability was assessed through PL analysis (Fig. 5f). gCN shows relatively high fluorescence intensity, suggesting rapid exciton recombination. Conversely, the binary composite shows spectra of relatively lower intensity, indicating an efficient segregation and transportation of photoinduced e^- and h^+. XPS

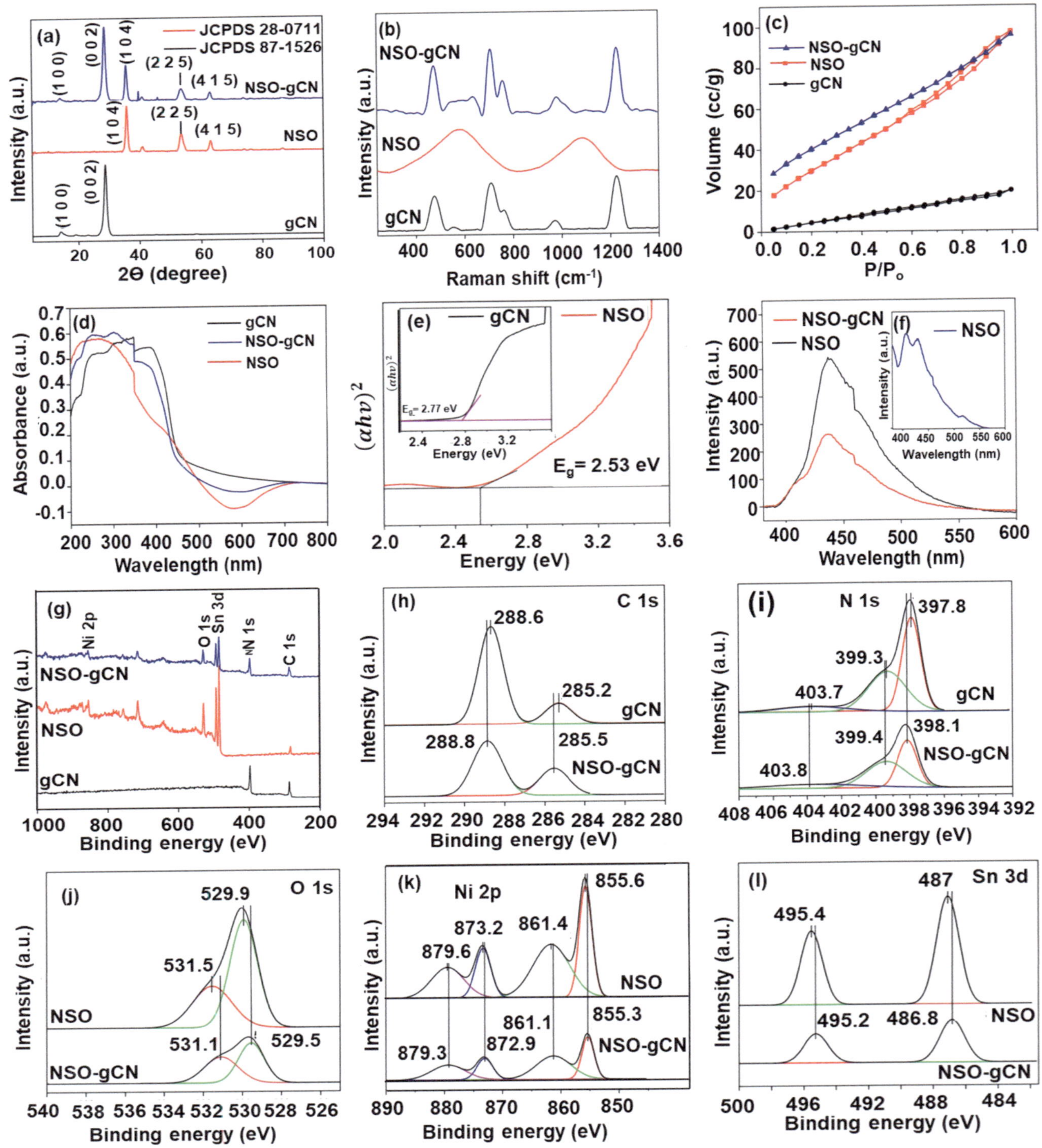

Fig. 5 **a** XRD and **b** Raman spectra, **c** N_2 adsorption–desorption isotherms, **d** UV-vis DRS spectra of gCN, NSO, and NSO-gCN; **e** Tauc plot of NSO and gCN; **f** PL spectra and **g** XPS survey scan, deconvoluted XPS spectra of **h** C 1s, **i** N 1s, **j** O 1s, **k** Ni 2p, and **l** Sn 3d of NSO, gCN, and NSO-gCN. Reprinted from Ref. [12], Copyright (2023), with permission from Elsevier

survey scan (Fig. 5g) confirmed the presence of C, N, O, Ni, and Sn in the composite. High-resolution deconvoluted C 1s spectrum (Fig. 5h) revealed distinct peaks from sp^2 C = C bonds (285.2 eV) and sp^2-bonded C (N–C = N) at 288.6 eV. Similarly, the N 1s spectrum (Fig. 5i) identified three distinct peaks at the binding energies of 397.8,

399.4, and 403.7 eV were attributed to C-N = C, N-$(C)_3$, and N–H bonds, respectively. Deconvolution of the O 1s spectrum for NSO (Fig. 5j) disclosed two oxygen states: lattice O^{2-} (529.9 eV) and surface-adsorbed oxygen (531.5 eV). The Ni 2p spectrum (Fig. 5k) confirmed the presence of Ni^{2+} ions through peaks at 873.2 (Ni $2p_{1/2}$) and 855.6 eV (Ni $2p_{3/2}$), accompanied by the characteristic satellite peaks. Finally, the Sn 3d spectrum (Fig. 5l) presented the peaks at the binding energies corresponding to 487.0 and 495.4 eV. The former and latter peaks corroborated to the presence of Sn $3d_{5/2}$ and Sn $3d_{3/2}$ oxidation states of Sn, respectively. NSO-gCN displayed each and every characteristic peak of its constituent components, confirming the successful formation of the heterojunction. However, a fascinating phenomenon emerged in the form of binding energy shifts. C and N peak positions were shifted toward high binding energies in NSO-gCN, while Ni, Sn, and O peaks shifted toward low binding energies. This intriguing pattern provides a compelling evidence of e^- migration from gCN to NSO, strongly supporting the S-scheme mediated charge transfer mechanism.

The XRD and FTIR analyses elucidated the crystalline structure and functional groups of gCN and FSO, confirming their successful incorporation into the FCN composite (Fig. 6a-b). The BET area data unveiled the porous characteristics of the synthesized materials (Fig. 6c). All samples exhibited a type IV isotherm characterized by an H3 hysteresis loop, reflecting the existence of slit-like mesopores within the structure. XPS provided detailed insights to the elemental composition and chemical states, revealing distinct peaks for C, N, O, Fe, and Sn in the materials (Fig. 6d). Deconvolution of the C and N spectra in gCN identified contributions from sp^2 bonds and various nitrogen configurations. The FCN composite exhibited notable binding energy shifts in the N, C, O, Fe, and Sn XPS spectra (Fig. 6e-i). The shifts suggest e^- transfer from gCN to FSO, offering strong validation for the proposed S-scheme-mediated charge transfer mechanism between FSO and gCN.

8 Materials Electrochemical Characterization

Understanding the charge carrier dynamics in FCN photocatalyst is crucial to elucidating the materials exceptional performance. Therefore, the Mott Schottky (MS) analysis was used to probe the band edge positions of the individual components of the heterojunction. FSO and gCN exhibited positive slopes in their respective MS plots (Fig. 7a, b), a signature characteristic of n-type semiconductors. This implies that their CBs are populated with readily available electrons, poised to engage in photocatalytic reactions. This finding assumes significance in the context of S-scheme heterojunctions. Ideally, such junctions are configured between two n-type semiconductors, promoting an efficient spatial segregation and directional transfer of photoinduced e^- and h^+. Consequently, the confirmation of n-type behavior in both FSO and gCN provides evidence for the effective formation of the desired S-scheme architecture in the FCN composite. The flat-band positions corresponding to 0.34 and −1.16 eV were determined for FSO and gCN, respectively, w.r.t. normal hydrogen electrode (NHE). EIS spectra were analyzed to compare the charge transfer resistance of the photocatalysts (Fig. 7c). The relatively smaller arc radius observed in the EIS spectra of FCN indicates reduced charge transfer resistance, emphasizing the improved segregation and transportation of the photogenerated e^- and h^+. Figure 7d illustrates the transient photocurrent response of the photocatalysts. The powdered photocatalysts were coated onto the conductive side of a fluorine-doped tin oxide electrode and subjected to five on–off cycles under the illumination of visible light. A significant rise in current density was noted upon light exposure, followed by a decline in the dark condition. The photocurrent density in the visible light was observed to be ~0.32, 0.11, and 0.025 $\mu A\ cm^{-2}$ for NSO-gCN, gCN, and NSO, respectively. The improved photocurrent density in NSO-gCN is ascribed to the generation of an S-scheme-mediated charge transfer heterojunction between gCN and NSO, facilitating efficient segregation and transportation of photogenerated h^+ and e^-, thereby reducing the charge carriers recombination rate.

9 Photocatalytic Activity

Photocatalytic H_2 evolution reaction was performed in a 150 mL glass reactor equipped with a cooling coil to maintain the reaction temperature at ~30 °C. The reactor housed 45 mL water containing 0.05 g of the photocatalyst. The mixture was stirred for 10 min and subsequently purged with N_2 for 30 min. Photodeposition of Pt NPs was achieved by irradiating the suspension with 250 W tungsten-halogen lamps for 30 min in the presence of 1% (w/w) H_2PtCl_6. After depositing Pt NPs, 5 mL of 10% (v/v) TEOA was added as a hole scavenger. The photoreactor was then purged with N_2 gas for an additional 0.5 h. Evolved H_2 was quantified hourly using the gas chromatography by injecting 1 mL gas samples through a syringe. Photocatalytic N_2 reduction to NH_3 occurred in a photoreactor. 280 mg photocatalyst was suspended in a 10% (v/v) methanol solution and stirred in the dark for 60 min. Under controlled temperature

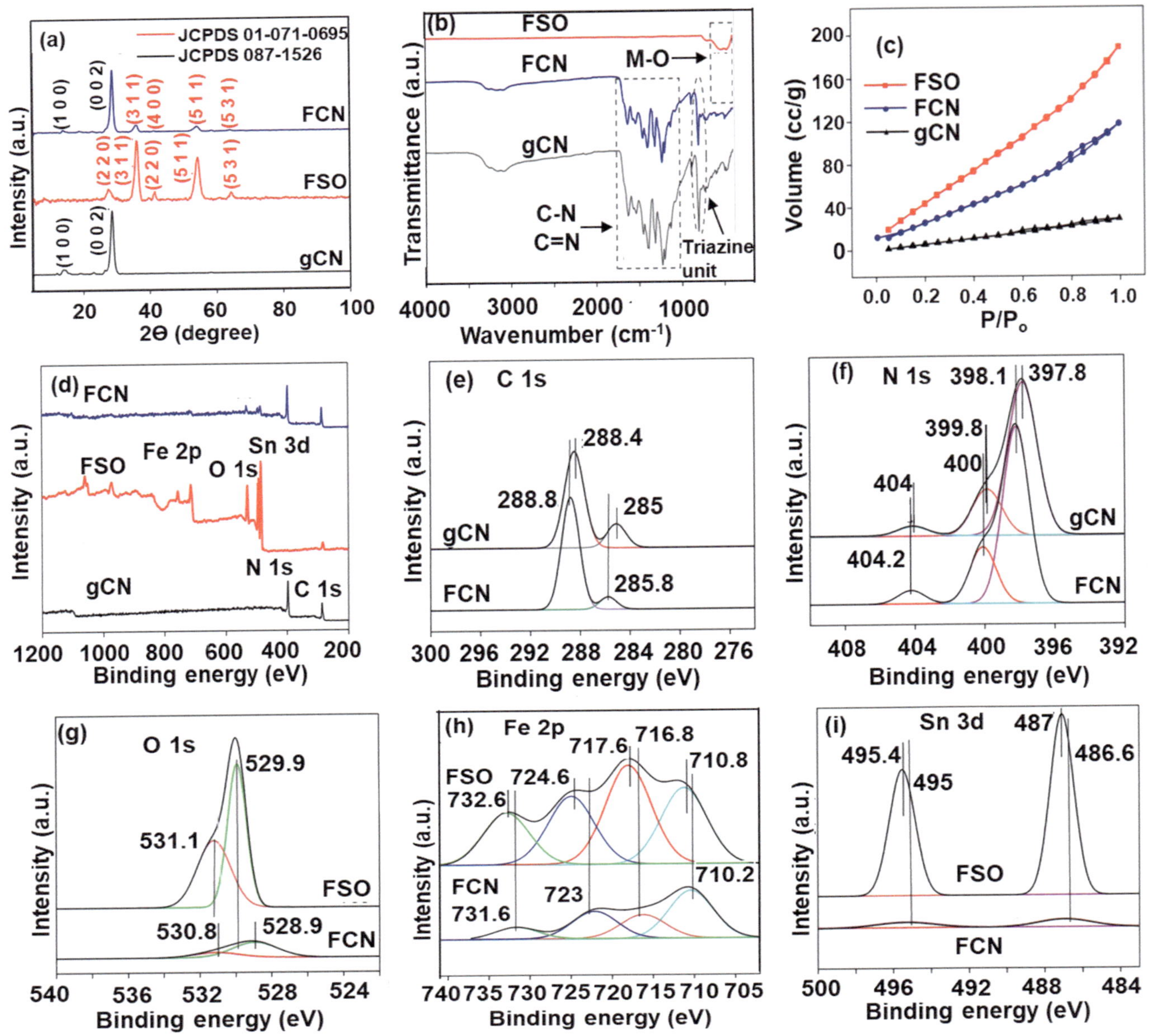

Fig. 6 **a** XRD, **b** FTIR, **c** N_2 adsorption–desorption isotherms of gCN, FSO, and FCN; XPS **d** survey spectra, deconvoluted spectra of **e** C 1s, **f** N 1s, **g** O 1s, **h** F e 2p, and **i** Sn 3d of gCN, FSO, and FCN. Reprinted from Ref. [13], Copyright (2024), with permission from Elsevier

(25–30 °C) and continuous N_2 purge (140 sccm), irradiation commenced for 4 h. Samples were collected at 30-min intervals and filtered using a 200 nm syringe filter. NH_4^+ ions concentration was determined using Nessler's reagent method. The absorbance at 420 nm was recorded using a UV–vis spectrophotometer, and the ammonia concentration was determined using a calibration curve generated from standard NH_4^+ solutions.

FCN exhibited a notable rise in the H_2 evolution rate, surpassing both FSO and gCN. While FSO exhibited a modest H_2 production rate of 9 μmol g^{-1} h^{-1}, gCN exhibited a remarkable enhancement at 322 μmol g^{-1} h^{-1}. However, the FSO-gCN composite significantly outperformed the rates, with an exceptional production rate of 1206 μmol g^{-1} h^{-1} (Fig. 8a, b). This represents a 3.7-fold enhancement compared to gCN and an astounding 133-fold increase over FSO, unequivocally demonstrating the synergistic efficacy of the S-scheme architecture in the FCN composite.

Figure 9a shows the variation in the NH_4^+ ion concentration with the time. NSO shows minimal NH_4^+ ion production (14 μmol g^{-1} h^{-1}), while gCN exhibits a moderate rate (154 μmol g^{-1} h^{-1}). NSO-gCN binary composite,

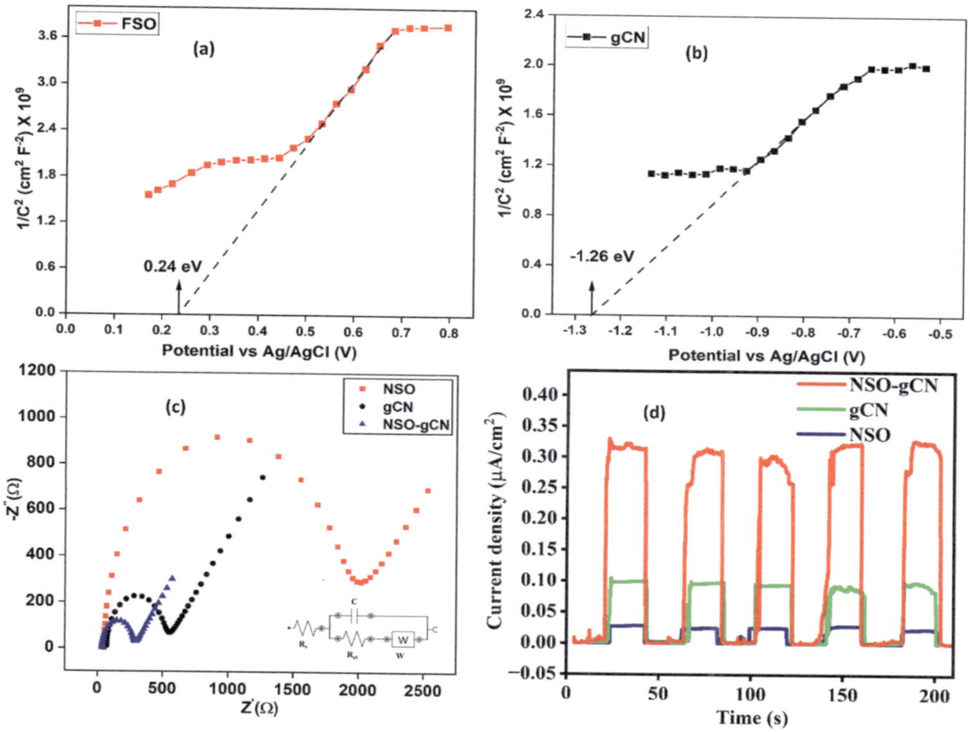

Fig. 7 MS plot of **a** FSO, **b** gCN, **c** EIS spectra, and **d** transient photocurrent response of gCN, NSO, and NSO-gCN. Reprinted from Refs. [12, 13, 15] respectively, Copyright (2023–2024), with permission from Elsevier

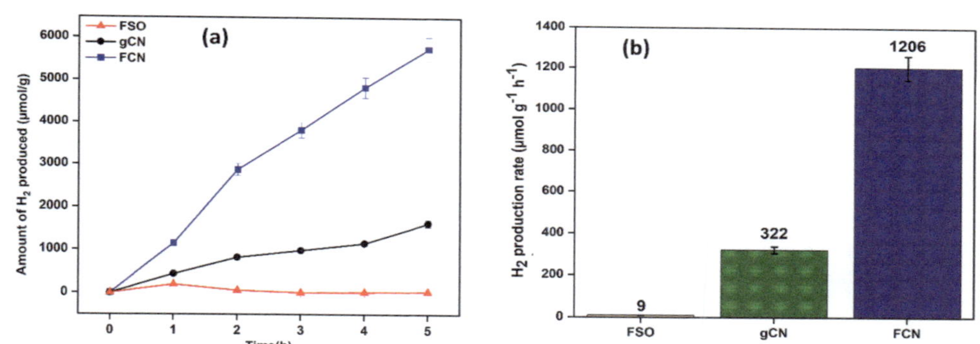

Fig. 8 **a** Time variation in the concentrations of generated H_2, and **b** the rate of photocatalytic H_2 production using FCN and its constituents (FSO and gCN). Reprinted from Ref. [13], Copyright (2024), with permission from Elsevier

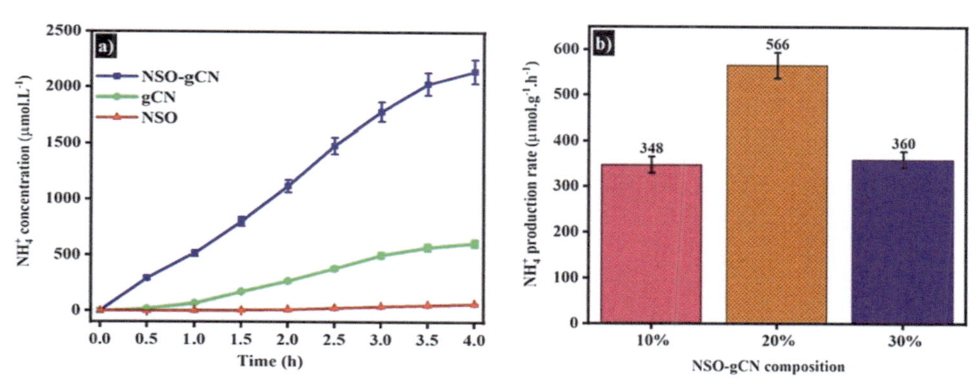

Fig. 9 **a** Time variation in the concentrations of photocatalytically generated NH_4^+ ions during the photoreduction of N_2, and **b** the rate of NH_4^+ ion production during photocatalytic N_2 fixation using different mass fractions of NSO (i.e., 10, 20, and 30%) in the NSO-gCN binary composite. Reprinted from Ref. [15], Copyright (2023), with permission from Elsevier

with 20% (w/w) NSO, achieves the highest photocatalytic activity, producing 566 μmol g^{-1} h^{-1}, which is 3.6 and 40 times higher than gCN and NSO, respectively. Figure 9b highlights the NH_4^+ ion generation rates for distinct mass fractions of NSO in NSO-gCN, revealing impressive NH_4^+ ion production rates ranging from 300 to 600 μmol g^{-1} h^{-1}.

10 Photocatalytic Mechanism for S-scheme Heterojunctions

Separate scavenging experiments were conducted using terephthalic acid (TA) and nitro blue tetrazolium (NBT) to confirm the formation of reactive oxygen species during the photocatalytic process. In the first experiment, a 0.025 × 10^{-3} M aqueous NBT solution was used to detect $^{\bullet}O_2^-$ radicals. Fig. 10a shows the UV–vis absorption spectra of NBT during the photocatalytic reaction. The reaction of NBT with $^{\bullet}O_2^-$ radicals lead to the formation of formazan blue, evidenced by the decreasing absorption peak at the wavelength of 258 nm. In the second experiment, a 0.5 × 10^{-3} M TA solution prepared in 2 × 10^{-3} M NaOH was utilized to detect $^{\bullet}OH$ radicals. The PL spectra in Figure 10b reveal an increase in intensity over time, attributed to the generation of TA-OH under photoirradiation, indicating the presence of $^{\bullet}OH$ radicals. Analysis of the band positions reveals that the CB of FSO is relatively lesser negative than the necessary potential for $O_2/^{\bullet}O_2^-$ (-0.33 eV vs NHE), thereby inhibiting the formation of $^{\bullet}O_2^-$ radicals. Similarly, the VB of gCN is lower than the required potential for $H_2O/^{\bullet}OH$ (1.99 eV vs. NHE), preventing gCN from independently generating $^{\bullet}OH$ radicals. The observations confirm that both radicals are formed only when FSO and gCN follow the S-scheme-mediated charge transfer mechanism.

It is true that S-scheme heterojunctions have emerged as promising candidates in harnessing solar energy for various energy and environmental applications. Their ability to facilitate efficient charge separation and utilization of photogenerated carriers makes them valuable in processes such as water splitting, nitrogen photofixation, pollutant degradation, and renewable energy production. Researchers now have focused on optimizing S-scheme configurations to enhance their overall performance. In this context, the exploration of dual S-scheme heterojunctions has taken center stage in photocatalysis research. This involves the integration of two distinct S-scheme configurations, creating a tandem structure that synergistically improves charge separation and promotes the efficient utilization of solar energy. Recent studies have shown that dual S-scheme heterojunctions offer enhanced capabilities in addressing the challenges associated with conventional heterojunctions, leading to superior outcomes in terms of energy conversion and environmental remediation. Recently, Zhao et al. [16] introduced a novel double S-scheme heterojunction, CeO_2/gCN/Bi_2O_4, as an effective strategy to enhance photocatalytic performance. This dual heterojunction significantly improves charge migration and achieves high redox potentials, offering distinct advantages in photocatalysis (Fig. 11). The CeO_2/gCN/Bi_2O $_4$composite demonstrated outstanding catalytic activity, achieving a 94% tetracycline removal efficiency and a 51% mineralization efficiency. Its tetracycline degradation rate constant (0.05811 min^{-1}) was 12.9, 7.8, 4.3, 2.2, and 3.1 times greater than that of gCN, CeO_2, Bi_2O_4, gCN/Bi_2O_4, and CeO_2/gCN, respectively. As researchers continue to delve into the intricacies of dual S-scheme heterojunctions, their potential for unlocking new avenues in sustainable energy and environmental applications becomes increasingly evident.

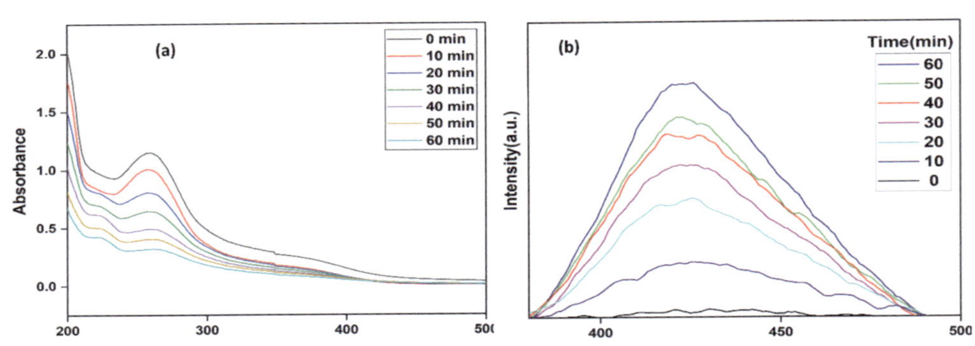

Fig. 10 **a** UV–vis absorption spectra of NBT and **b** PL spectra of TA with irradiation time. Reprinted from Ref. [13], Copyright (2023), with permission from Elsevier

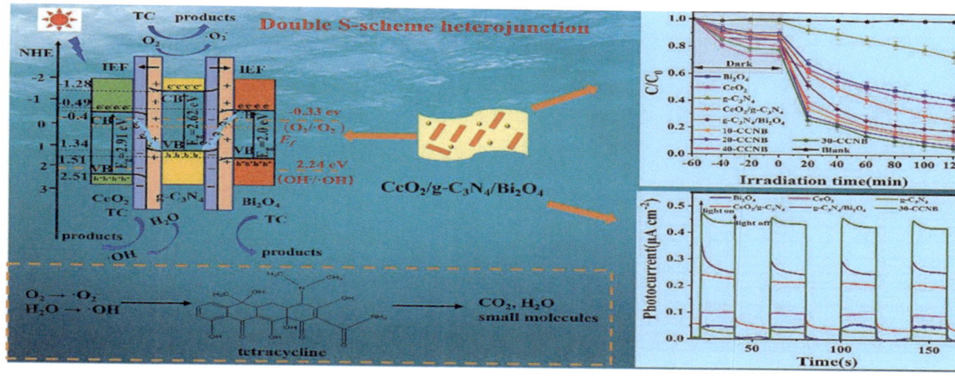

Fig. 11 Schematic representation of photocatalytic tetracycline degradation by the dual S-scheme CeO_2/g-C_3N_4/Bi_2O_4 heterojunction. Reprinted from Ref. [16], Copyright (2024), with permission from Elsevier

Acknowledgements Authors express their gratitude to the Centre for Environmental Science and Engineering, IIT Kanpur for providing the research facilities. The contributions of both past and present graduate students, who have studied various aspects of S-scheme heterojunctions for photocatalytic applications, are greatly acknowledged.

References

1. Zhang L, Zhang J, Yu H, Yu J (2022) Emerging S-scheme photocatalyst. Adv Mater 34:2107668. https://doi.org/10.1002/adma.202107668
2. Xu Q, Zhang L, Cheng B, Fan J, Yu J (2020) S-scheme heterojunction photocatalyst. Chem 6:1543–59. https://doi.org/10.1016/j.chempr.2020.06.010
3. Hao P, Chen Z, Yan Y, Shi W, Guo F (2024) Recent advances, application and prospect in g-C3N4-based S-scheme heterojunction photocatalysts. Sep Purif Technol 330:125302. https://doi.org/10.1016/j.seppur.2023.125302
4. Zhu B, Sun J, Zhao Y, Zhang L, Yu J (2023) Construction of 2D S-scheme heterojunction photocatalyst. Adv Mater. https://doi.org/10.1002/adma.202310600
5. Li F, Zhu G, Jiang J, Yang L, Deng F, Arramel, et al (2024) A review of updated S-scheme heterojunction photocatalysts. J Mater Sci Technol 177:142–80. https://doi.org/10.1016/j.jmst.2023.08.038
6. Rana S, Kumar A, Sharma G, Dhiman P, García-Penas A, Stadler FJ (2023) Recent advances in perovskite-based Z-scheme and S-scheme heterojunctions for photocatalytic CO_2 reduction. Chemosphere 339:139765. https://doi.org/10.1016/j.chemosphere.2023.139765
7. Nie C, Wang X, Lu P, Zhu Y, Li X, Tang H (2024) Advancements in S-scheme heterojunction materials for photocatalytic environmental remediation. J Mater Sci Technol 169:182–198. https://doi.org/10.1016/j.jmst.2023.06.011
8. Mishra R, Bera S, Chatterjee R, Banerjee S, Bhattacharya S, Biswas A et al (2022) A review on Z/S—scheme heterojunction for photocatalytic applications based on metal halide perovskite materials. Appl Surf Sci Adv 9:100241. https://doi.org/10.1016/j.apsadv.2022.100241
9. Dharani S, Vadivel S, Gnanasekaran L, Rajendran S (2023) S-scheme heterojunction photocatalysts for hydrogen production: current progress and future prospects. Fuel 349:128688. https://doi.org/10.1016/j.fuel.2023.128688
10. Gupta R, Alam U, Kumar R, Kumar S, Verma N (2025) S-scheme-mediated $NiCo_2S_4$–P–*g*-C_3N_4 heterojunction for selective nitrate reduction to ammonia. ACS Applied Engineering materilas. https://doi.org/10.1021/acsaenm.4c00707
11. Yue L, Zhang J, Zeng Z, Zhao C, Hu X, Zhao L et al (2023) In situ fabrication of an S-scheme $NaNbO_3$ /Bi_2 O_2 CO_3 heterojunction for enhanced performance in photocatalytic nitrogen fixation. Langmuir 39:13267–13278. https://doi.org/10.1021/acs.langmuir.3c01725
12. Pandey A, Alam U, Gupta A, Shim J-J, Verma N (2024) S-scheme heterojunction-mediated hydrogen production over the graphitic carbon nitride-anchored nickel stannate perovskite. Fuel 355:129538. https://doi.org/10.1016/j.fuel.2023.129538
13. Gupta A, Alam U, Verma N (2024) Efficient spatial charge separation in the Ruddlesden-Popper phase-based S-scheme Fe_2SnO_4-g-C_3N_4 heterojunction for visible light-induced H2 generation. Int J Hydrogen Energy 58:69–78. https://doi.org/10.1016/j.ijhydene.2024.01.102
14. Alam U, Pandey A, Verma N (2023) An anthraquinone-integrated S-scheme-based $NiTiO_3$-g-C_3N_4 composite with enhanced hydrogen production activity. Int J Hydrogen Energy 48:2532–2541. https://doi.org/10.1016/J.IJHYDENE.2022.10.151
15. Gupta R, Alam U, Verma N (2024) Efficient photofixation of nitrogen to ammonia over an S-scheme-based $NiSnO_3$-g-C_3N_4 heterojunction. Chem Eng J 479:147644. https://doi.org/10.1016/j.cej.2023.147644
16. Zhao S, Jiang J, Zhang C, Chen F, Song Y, Tang Y (2024) Construction of a novel double S-scheme heterojunction CeO_2/g-C_3N_4/Bi_2O_4 for significantly boosted degradation of tetracycline: Insight into the dual charge transfer mode. Chem Eng J 479:147333. https://doi.org/10.1016/j.cej.2023.147333

Molten Catalytic Systems for CO_2-Free H_2 Production from CH_4: A Perspective from Ab Initio Simulations

Nikil Surya Rajaprabhu, Aditya Goyal, and Vishal Agarwal

1 Introduction

Hydrogen is an important feedstock to many industrial chemical processes [5, 6] and, as the world is moving toward decarbonization, it is being promoted as the "green" fuel of the future. Hydrogen also has a global socio-economic footprint because it is the feedstock for synthesis of NH_3, which serves as the biggest source of producing nitrogenous fertilizers [18]. Currently, most of the hydrogen is produced from steam methane reforming (SMR) [3, 7, 9] and coal gasification (CG) [9, 29, 58]; the former being the most economical and acceptable process for H_2 generation. SMR uses steam at harsh conditions, typically at 800–900 °C temperatures and 14–20 bar pressures, to strip off the hydrogen from methane using a metal-based catalyst with an overall reaction of

$$\text{SMR}: CH_4 + H_2O \leftrightharpoons CO + 3H_2, \qquad \Delta H = 206\ \text{kJ/mol}. \tag{1.1}$$

On a metal surface the overall process of SMR is proposed to occur via several elementary steps, which are shown in Fig. 1. A comprehensive recent study computed the barriers for all these steps on Ni(111) and Pt(111) using density functional theory (DFT) calculations [32]. Within the theoretical error, the apparent activation energy obtained by performing microkinetic analysis was a good match with that obtained from experiments. SMR is a highly endothermic process and produces carbon-monooxide along with three moles of hydrogen for every mole of CH_4 consumed. An additional mole of hydrogen is produced by catalytically reacting carbon-monooxide with steam via water gas shift (WGS) reaction.

$$\text{WGS}: CO + H_2O \rightarrow CO_2 + H_2 \qquad \Delta H = -41.2\ \text{kJ/mol}. \tag{1.2}$$

The overall process produces *four* moles of H_2 from *one* mole of methane supplied. The carbon footprint comes from the CO_2 produced and the fossil fuel burnt to supply energy for the process. It is estimated that 11–12 tons of CO_2 is discharged into environment for every ton of H_2 produced during the SMR [36, 40]. This is problematic for the envisioned "green" future, which is expected to run on zero-carbon technologies. SMR can, however, be coupled with carbon capture and sequestration (CSS) [1, 30] to eliminate or reduce CO_2 emissions [52], which is an additional effort and a costly proposition. Alternatively, methane pyrolysis is a potential option for CO_2-free production of hydrogen.

In a pyrolysis process, methane is cracked to carbon and hydrogen at high temperatures.

$$\text{Methane pyrolysis}: CH_4 \rightarrow C + 2H_2, \qquad \Delta H = 75.6\ \text{kJ/mol}. \tag{1.3}$$

The energy required to break a C–H bond in methane is more than 400 kJ/mol [16, 54], and therefore efficient decomposition of methane and hydrogen in a non-catalytic pyrolysis requires temperatures which are in excess of 1000 °C. The reaction in gas-phase proceeds via a combination of initiation, propagation, and termination steps—the rate-determining step of which is the breaking of the CH_3–H bond [54]. Methane pyrolysis, on the other hand, can be accomplished at lower temperatures using metal catalysts such as Ni or Fe dispersed on a highly porous support. The reaction is, however, equilibrium-limited, and therefore requires high temperatures and low pressures to achieve very high conversions of methane. A plot of conversion *versus* temperature is shown in Fig. 2. At equilibrium, under 1 bar pressure, approximately 92% conversion of methane is achieved at 800°C and 98% conversions at 1000°C. High

N. S. Rajaprabhu and A. Goyal—N.S.R. and A.G. have contributed equally.

N. S. Rajaprabhu · A. Goyal · V. Agarwal (✉)
Department of Chemical Engineering, Indian Institute of Technology Kanpur, Kanpur, India
e-mail: vagarwal@iitk.ac.in

V. Agarwal
Department of Sustainable Energy Engineering, Indian Institute of Technology Kanpur, Kanpur, India

S. Narayanan et al. (eds.), *Materials, Devices and Systems for Sustainability*, IITK Directions 8,
https://doi.org/10.1007/978-981-96-7295-0_6

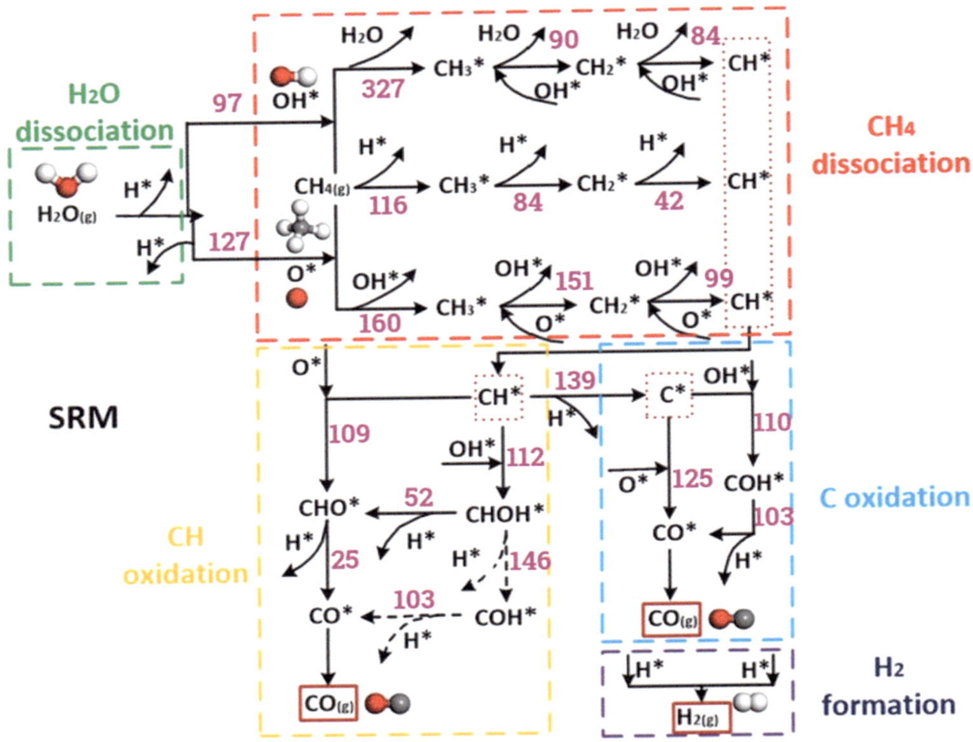

Fig. 1 Proposed elementary reaction pathways of steam methane reforming on the surface of a metal catalyst. The entire network is categorized into five distinct segments: H_2O dissociation, CH_4 dissociation, CH oxidation, C oxidation, and H_2 formation. The numerical values indicate the activation barriers (kJ/mol) on a pristine Ni(111) surface computed using DFT. Adapted from Niu et al. [32]

temperatures comes with its own set of problems. The catalyst dispersed on porous support may frittage or sinter at high temperatures, which reduces its activity dramatically [19, 35, 48]. The main problem with solid catalyst, however, is the rapid deactivation as a result of carbon accumulation on the surface [2]. Although gasification or combustion can prolong the catalyst's life, this approach goes against our objective of generating "green" H_2 [15].

To avoid deactivation, methane pyrolysis can be carried in a reactor filled with molten catalyst. Methane bubble is introduced at the bottom of the reactor. As the bubble rises in a bath of molten catalyst, methane is cracked at the liquid surface encompassing the bubble to carbon and hydrogen. A sparger can be used to introduce smaller bubbles for increasing residence time and effective contact area and hence the conversions [26, 30]. The insoluble carbon is carried with the bubble to the top of the molten phase, where it accumulates and can be continuously removed [11, 45]. A schematic figure of molten reactor is shown in Fig. 3. There are distinct advantages of using a molten catalytic systems. First, one does not have to go through a sophisticated process of preparing and characterizing the catalyst. Second, and most important, the carbonaceous compounds formed are separated automatically. The active sites in the bulk catalyst are, therefore, free from fouling and in principle can be used infinitely. For a methane-fired pyrolysis reactor, there is a reduction of approximately 85% of CO_2 emissions per H_2 produced in comparison to a steam reformer [36, 40]. If the fraction of hydrogen produced

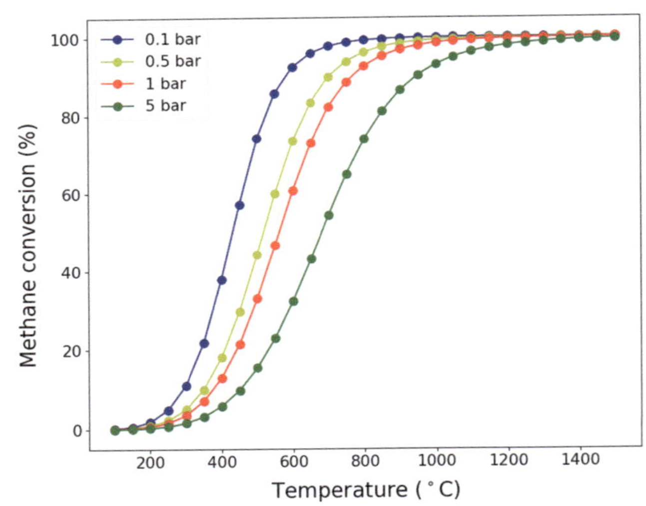

Fig. 2 Equilibrium conversion of methane at different temperatures and pressures calculated using Peng–Robinson thermodynamic model in Aspen plus V 8.8 (Aspen Tech, Bedford, USA)

in the process is used for its heat requirement, there is no CO_2 emissions. Although commercial implementation of molten catalyst reactor has still not been achieved, however, several cost estimates place this technology at par with SMR if a CO_2 abatement tax is imposed [40].

Molten catalysts offer great promise for future applications to wide variety of synthesis, especially where coking is

Table 1 Summary of molten catalysts studied in literature for methane pyrolysis to carbon and hydrogen. The activation barriers for those which have been reported are also given. A rising bubble may be catalyzed by vaporized metals or metal halides, which are given at the end of the table. Values in parenthesis in the second column represent the mol% of a particular component used in the mixture. "E" and "T" enclosed in the parenthesis in the third column signify experimental and theoretical, respectively

System	Catalyst	Apparent activation energy (kJ/mol)	References
Non-catalytic	–	423 (E)	[54]
Catalytic			
Molten metal	Fe	–	[57, 63]
	Al	–	[33]
	Mg	–	[33, 65]
	Sn	–	[20, 21, 33, 44, 46, 50, 51, 56, 57, 64]
	Sn	205 (E)	[66]
	Pb	–	[33, 51, 64]
	Zn	–	[14, 33]
	Sn/SiC	–	[51]
	Cu	–	[61]
	Ag	–	[64]
	Bi	310 (E)	[64]
	Bi	–	[10]
	In	–	[64]
	In/SiO_2 support	170 (E), 223 (T)	[31]
	Ga	–	[45, 64]
	Te	166 (E), 139 (T)	[67]
	Na	206–240 (T)	[47]
Molten metal alloys	Fe-Mn	–	[63]
	Fe-Ni	–	[63]
	Al-Cu	–	[33]
	Al-Mg	–	[33]
	Al-Ni†	–	[33]
	Sn-Pb	–	[33]
	Cu-Pb	–	[28, 33]
	Ni(27)-Bi(73)	208 (E)	[64]
	Ni-Bi†	–	[10, 13, 28]
	Cu(17)-Sn(83)	–	[64]
	Cu(8.9)-Sn(91.1)	207 (E)	[66]
	Cu(44.5)-Sn(55.5)	250 (E)	[66]
	Pt(17)-Sn(83)	–	[28, 64]
	Pt-Bi†	–	[64]
	Ni-In†	–	[28, 64]
	Ni(65)-In(35)	322±22 (E)	[39]
	Ni-Sn†	–	[13, 64]
	Ni(9.6)-Sn(90.4)	235 (E)	[66]
	Ni(33)-Sn(67)	355±9 (E)	[55]
	Ni-Ga†	–	[14, 64]
	Ni(17)-Pb(83)	–	[64]
	Ni(27)-Au(73)	–	[64]
	Ga(77)-In(23)	–	[42]
	Ga(68.6)-In(21.4)-Sn(9.9)	–	[42]
	Cu(45)-Bi(55)	222±20 (E), 245–255 (T)	[38]
	Cu-Bi†	–	[10, 28]

(continued)

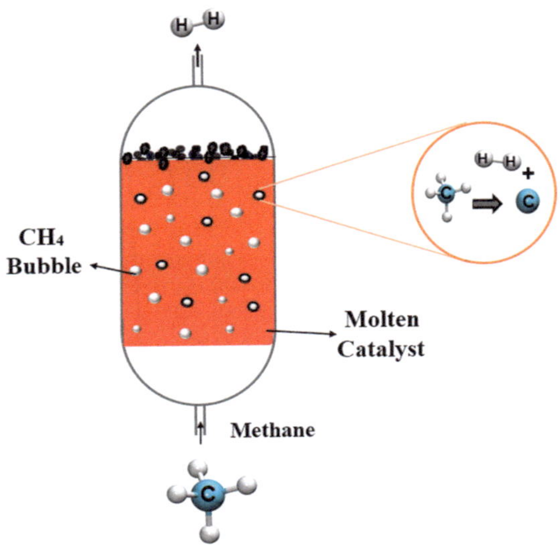

Fig. 3 Methane pyrolysis in a bubble column reactor. Carbon nucleates in the rising bubble and separates on the top

a difficult problem to handle. In recent times, several molten catalysts based on metals, metal halides, and mixtures of them have been tested. A summary of these is presented in Table 1 along with apparent activation energies measured through experiments and/or computed using quantum mechanical calculations. At high temperatures, a rising bubble will also have gas-phase metals or metal halides, which may catalyze methane pyrolysis. We also present computational work done to understand vapor-catalyzed methane pyrolysis in Table 1. There are several excellent reviews [8, 27, 30, 34, 68, 69] which have been published on molten catalysts, however, the focus has been the experimental work performed in this field. With the advent of efficient implementation of first-principles calculations along with tremendous increase in the computational resources, it is now possible to qualitatively understand and predict properties of novel catalytically materials. Understanding electronic attributes, encompassing electron distribution and energy levels, assists in anticipating the catalyst's proficiency in facilitating electron transfer during methane pyrolysis. With this is in mind, we present in this chapter a comprehensive overview of *ab initio* simulations done to quantify the nature of catalytic sites in molten systems with the focus on methane pyrolysis.

2 Molten Metals and Alloys

Liquid metals and alloys have unique characteristics that make them very customizable and adaptable to a variety of applications [12, 68]. They exhibit properties that overcome the drawbacks of conventional heterogeneous catalysis, which is often linked with easier product separation and catalyst recovery [12, 49, 60, 69]. The synthesis of hydrogen by methane pyrolysis is one typical use for molten metal (or alloy) catalysts.

The first report of molten metal catalyst used for methane pyrolysis dates back to 1931 [63]. The authors demonstrated that methane pyrolysis can be carried out at 1200–1300 °C in molten Fe, which was alloyed with Mn or Ni to increase its activity. The main advantage was separation of the carbon, which was deposited at the top of the molten catalyst. Although intermittent efforts to develop molten metal/alloy catalyst for methane pyrolysis were attempted, molten catalyst technology has recently come into focus as our society is moving toward decarbonization. We summarize to the best of our efforts, a list of molten metal/alloys attempted for methane pyrolysis in Table 1. In recent times, state-of-the art quantum chemical simulations have also been performed to understand the nature of active sites in these molten systems, which is the focus of the current chapter.

In a pioneer work, Upham et al. [64] systematically examined the catalytic effect of alloying low-melting (but inactive) metals such as indium, gallium, tin, and lead with Ni and Pt. Among these bimetallic catalysts, Bi alloyed with 27mol% Ni ($Ni_{0.27}Bi_{0.73}$) proved to be the most potent liquid catalyst, exhibiting an enhanced rate of methane conversion with an effective activation barrier of 208 kJ/mol. The maximum Ni content was limited by the melting temperature of the alloy. It is expected that higher Ni percentage would have enhanced the rates even further. The atomic nature of the catalyst was investigated by performing AIMD simulations based on density function theory at constant temperature. The simulation cell consisted of 82 atoms of low-melting metal with a vacuum layer on top of the molten catalyst. Periodicity was used in all three directions, and therefore the system mimics an infinite thin film of liquid. Several interesting insights ensued based on these simulations. First, the active metal such as Pt or Ni is present as negative ion in the melt, which is surrounded by a shell of positive inactive metal. Figure 4a shows the plot of Bader charges on Pt atom and the total charge on surrounding Sn atoms as a function of simulation time in a molten Pt-Sn system. Second, the additional electron on the active metal occupies the d-orbital (see Fig. 4b)—the shape of which dynamically fluctuates as the configuration of the "solvent" metal around the active metal changes. Third, the active metals are present as negatively charged atomic species. To verify this, a study involving evolution of Pt clusters (Pt_2 and Pt_8) placed in molten Sn was done. Interestingly, Pt_2 and Pt_8 cluster in molten Sn dissociated within order of picoseconds, resulting in the formation of Pt^{-1} ions as shown in Fig. 4c. This observation is extremely surprising because in a liquid the dissociation reactions are thwarted by what is known as "cage-effect". It was suggested that the electrons from the surrounding solvent atoms rush to occupy the anti-bonding orbitals of the active metals, which rapidly dissociates the

Table 1 (continued)

System	Catalyst	Apparent activation energy (kJ/mol)	References
	Ni(27)-Te(73)	193 (E)	[67]
	Ni(50)-Te(50)	223 (E)	[67]
	Cu-Ga†	–	[14, 28]
	Fe-Ga†	–	[14, 28]
	Cu-In†	–	[28]
	Cu(60)-Te(40)	–	[28]
	Pd(20)-Ga(80)	–	[28]
	Co(9.5)-Sn(90.5)	–	[28]
	Mo-Bi	–	[10]
	Zn-Bi	–	[10]
	Zn-Mo-Bi	–	[10]
	Cu-Mo-Bi	–	[10]
	Ni-W-Bi	–	[10]
	Ni(7.6)-Mo(2.6)-Bi(89.8)*	81 (E)	[10]
Molten salt	NaCl	–	[57]
	NaF	–	[57]
	$MnCl_2$	178±17 (E)	[23]
	KCl	288±16 (E), 317 (T)	[37]
	NaCl	309±14 (E)	[37]
	KBr	310±22 (E)	[37]
	NaBr	295±16 (E)	[37]
	KCl	236±9 (E)	[41]
	NaCl	231±21 (E)	[41]
	KBr	224±2 (E)	[41]
	NaBr	278±13 (E)	[41]
	$MnCl_2$	172±14 (E)	[4]
	LiCl	–	[28]
	LiBr	–	[28]
	$CaCl_2$	–	[53]
	$MgCl_2$	–	[53]
	$ZnCl_2$	–	[53]
Molten salt mixtures	$MnCl_2$(67)-KCl(33)	161±9 (E)	[23]
	$MnCl_2$(50)-KCl(50)	153±16 (E)	[23]
	NaCl(50.5)-KCl(49.5)	301±7 (E)	[22]
	NaKCl(97.5)-$FeCl_3$(2.5)	171±5 (E)	[22]
	NaBr(48.7)-KBr(51.3)	247±7 (E)	[41]
	NaBr(48.7)-KBr(51.3)	236±6 (E)	[43]
	$MnCl_2$(50)-NaCl(50)	184±13 (E)	[4]
	$MnCl_2$(50)-LiCl(50)	196±17 (E)	[4]
	$MnCl_2$(50)-$MgCl_2$(50)	223±27 (E)	[4]
	$MnCl_2$(50)-$CaCl_2$(50)	210±25 (E)	[4]
	KCl(99)-$FeCl_3$(1)	128-194 (T)	[17]
	KCl(80)-$MgCl_2$(20)	240 (E)	[53]
	KCl-$MgCl_2^{\dagger}$	–	[53]
	KCl(80)-$CaCl_2$(20)	275 (E)	[53]
	KCl-$CaCl_2^{\dagger}$	–	[53]
	KCl(80)-$CrCl_2$(20)	198 (E)	[53]
	KCl(80)-$FeCl_2$(20)	183 (E)	[53]

(continued)

Table 1 (continued)

System	Catalyst	Apparent activation energy (kJ/mol)	References
	KCl(80)-$ZnCl_2$(20)	364 (E)	[53]
	KCl-$ZnCl_2^{\dagger}$	–	[53]
	KCl(80)-$CoCl_2$(20)*	265 (E)	[53]
	KCl(80)-$NiCl_2$(20)*	98 (E)	[53]
	KCl(80)-$CuCl_2$(20)*	223 (E)	[53]
	KCl(80)-AgCl(20)*	243 (E)	[53]
Metal nanoparticles-molten salt	MnCo/Al-NaBr-KBr	195±5 (E)	[43]
	$MnCo_2$/Al-NaBr-KBr	176±6 (E)	[43]
	Ni-NaBr	–	[62]
Gaseous metal	Te	178 (E), 239 (T)	[67]
	Na	296 (T)	[47]
	Pb	–	[64]
Gaseous metal halides	$ZnCl_2$	227 (T)	[16]
	$CuCl_2$	122 (T)	[16]
	$CoCl_2$	177 (T)	[16]
	$AlCl_3$	170 (T)	[16]
	$FeCl_2$	192 (T)	[16]
	$NiCl_2$	161 (T)	[16]

*The doped metal halide reduces to their respective metals for these cases. For example, $NiCl_2$ reduces to Ni metal in a KCl(80)-$NiCl_2$(20) molten system during methane pyrolysis

†Several different compositions were tried

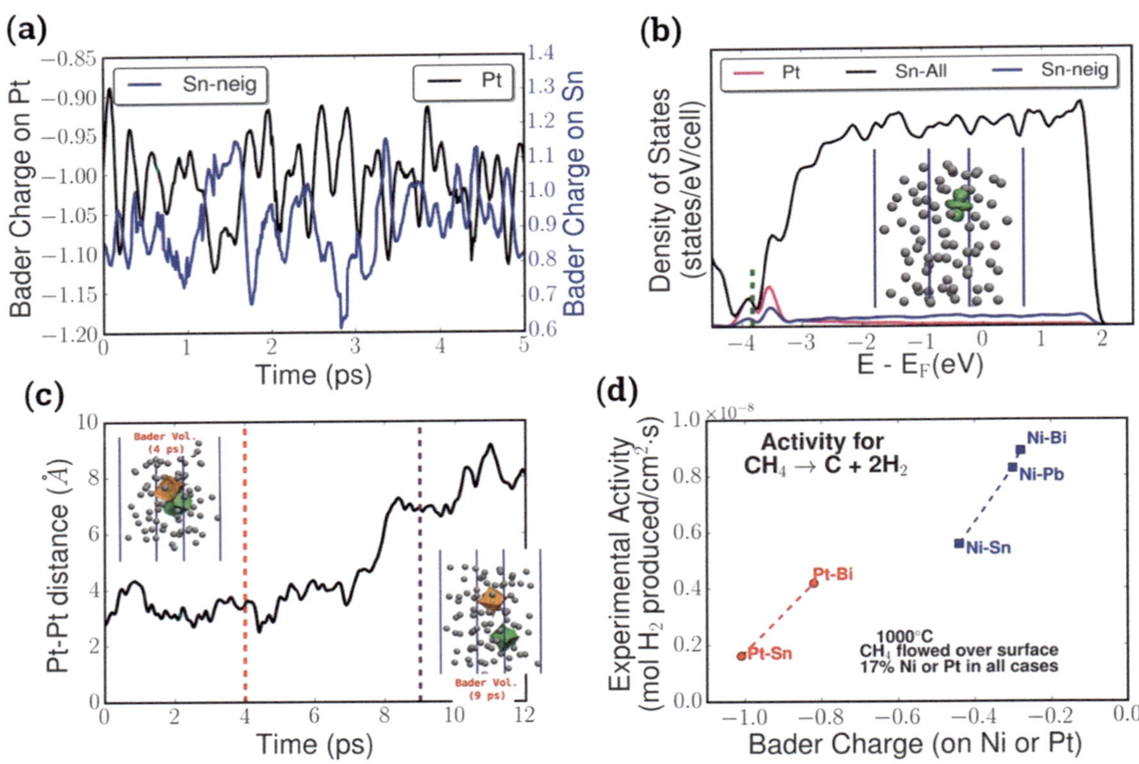

Fig. 4 **a** Charges on Pt and the total charge on Sn atoms in the first shell surrounding Pt as a function of time. **b** Projected density of states on selected atoms. **c** Inter-atomic Pt-Pt distance as a function of time. **d** Activity of methane at 1000 °C plotted as a function of Bader charges on supposedly active elements. Adapted from Upham et al. [64]

metal clusters as in Coulombic explosion. Fourth, the Bader charge on the active metal correlated well with the experimental activity as shown in Fig. 4d—lesser the negative charge on the active metal, more is the activity.

Very recently, it was shown that the activity of Ni-Bi alloy can be further enhanced by adding Mo as the third metal [10]. The experimental barriers were reported to be 81 kJ/mol, which is much lower than that observed for Ni-Bi alloy. It was suggested that Mo regulates the activity by modulating the interaction of Ni and Bi atoms. Ab initio molecular dynamic simulations performed by the authors suggested a slight reduction in negative charge on the Ni atom and is not able to account for dramatic reduction in barriers for methane pyrolysis. A more thorough simulation studies can potentially elucidate the electronic and structural properties of the active metal atom in the NiMo-Bi ternary. A particularly interesting question to ask is: whether Mo modified by the presence of surrounding Bi and Ni acts as the active site for catalysis?

In a subsequent study of molten Ni-Bi catalyst, the same group [38] studied the catalytic properties of alloying Cu and Bi. Surfaces made of copper and bismuth alloy had slightly better activity for methane conversion than molten Ni(27mol%)-Bi(73 mol%), and, additionally, copper is more cost-effective than nickel. Both Cu and Bi are inactive for methane pyrolysis, however, on alloying the authors found a considerable enhancement in the catalytic activity. The authors also found that the alloy whose surface was more enriched with Bi was more active. To understand these surprising findings, the authors simulated molten Cu-Bi at different alloying percentages ($Cu_{10}Bi_{90}$, $Cu_{20}Bi_{80}$, and $Cu_{45}Bi_{55}$) through constant temperature AIMD. In these simulations, an infinite liquid slab was used which consisted of a vacuum–liquid interface on either side. The number density of Cu and Bi in the molten alloy model system was computed as a function of height, as shown in Fig. 5b. The density profile shows that Bi atoms segregate on the surface across all bulk concentrations, and, notably, the outermost layers are mainly composed of Bi. This observation aligns with the experimental findings and is in accordance with the principle that a metal with high surface tension (Cu) tends to preferentially separate away from the surface over the low-surface-tension component (Bi). The system having more Bi is less compacted, which is evident from the increase in the height of liquid slab.

To quantify the electronic nature of the atoms in the molten Cu-Bi alloys, the authors computed the mean Bader charge on Cu and Bi [38]. Figure 5c shows the temporal changes in the mean Bader charge on Cu and Bi. In comparison to gas-phase atoms, Bi is positively charged and Cu is negatively charged. As the concentration of Bi increases in the bulk, Bi becomes slightly less positive, as expected. Bi is less positively charged than in the bulk, while Cu is less negatively charged (see Fig. 5d).

The dissociation of CH_3-H bond is commonly recognized as the step that predominantly determines the reaction rate in methane pyrolysis. Performing simulations to investigate the catalysis over a molten catalyst is a computational daunting task due to the dynamic nature of the catalyst's surface. Therefore, to understand qualitative nature of the methane activation barriers, the authors chose a random configuration from an extended molecular dynamics (MD) run. And created a glassy solid by fixing the lower 80% of the slab to establish a structure resembling that of a liquid. Subsequently, binding configurations of CH_3 and H were explored by placing these species at various locations on the surface of Cu_{45}-Bi_{55}. Out of the several possibilities, H and CH_3 preferred to be bonded to Bi atoms rather than Cu (see Fig. 5e). Transition-state calculations were performed on the two most stable product configurations to calculate the activation barrier for methane pyrolysis. In the first stable product configuration, both CH_3 and H were placed on the bridging site of multiple Bi atoms on the surface, while in the second, CH_3 was positioned on single Bi, and H was situated between multiple Cu atoms. The activation energies for these two paths were found to be similar, at 244 and 251 kJ/mol. The experimentally determined apparent activation barrier was determined to be 222 kJ/mol, and, within the margin of theoretical error, is consistent with that determined from DFT calculations [38].

Among the low-melting metals, tellurium has a very high electron affinity, and therefore Zeng et al. [67] explored molten tellurium as a catalyst for methane pyrolysis. Tellurium also has very high vapor pressures and boils at 988 $^\circ$C, and therefore presents a possibility of gas-phase catalysis. It was found that tellurium exhibited catalytic efficacy in both liquid and vapor phases for methane decomposition within the range of reaction temperature. A structure similar to that of a liquid is created by fixing 70–80% of bottom part of the slab. The barrier involved in the CH_3-H dissociation step on a "glassy" solid consisting of tellurium atoms was computed to determine the activation barrier for the dissociation of methane on molten tellurium. Several product configurations were obtained, which are shown in Fig. 6a). Depending on whether the species are attached to single or multiple Te atoms on the surface, the product configurations were reported to have a dispersion of roughly 0.6 eV (see Fig. 6a). The most stable product configuration is the one in which these species are bound to a single Te atom on the surface [67]. The activation barrier for methane dissociation was computed using the most stable product configuration. The computational approach was applied to investigate a sequence wherein the initial state involves methane adsorption on the surface of liq Te, progressing through the dissociation process to form the most stable configuration depicted in Fig. 6b–e. The computed activation barriers were found to be less for molten Te in comparison to vapor Te, in agreement with experiments.

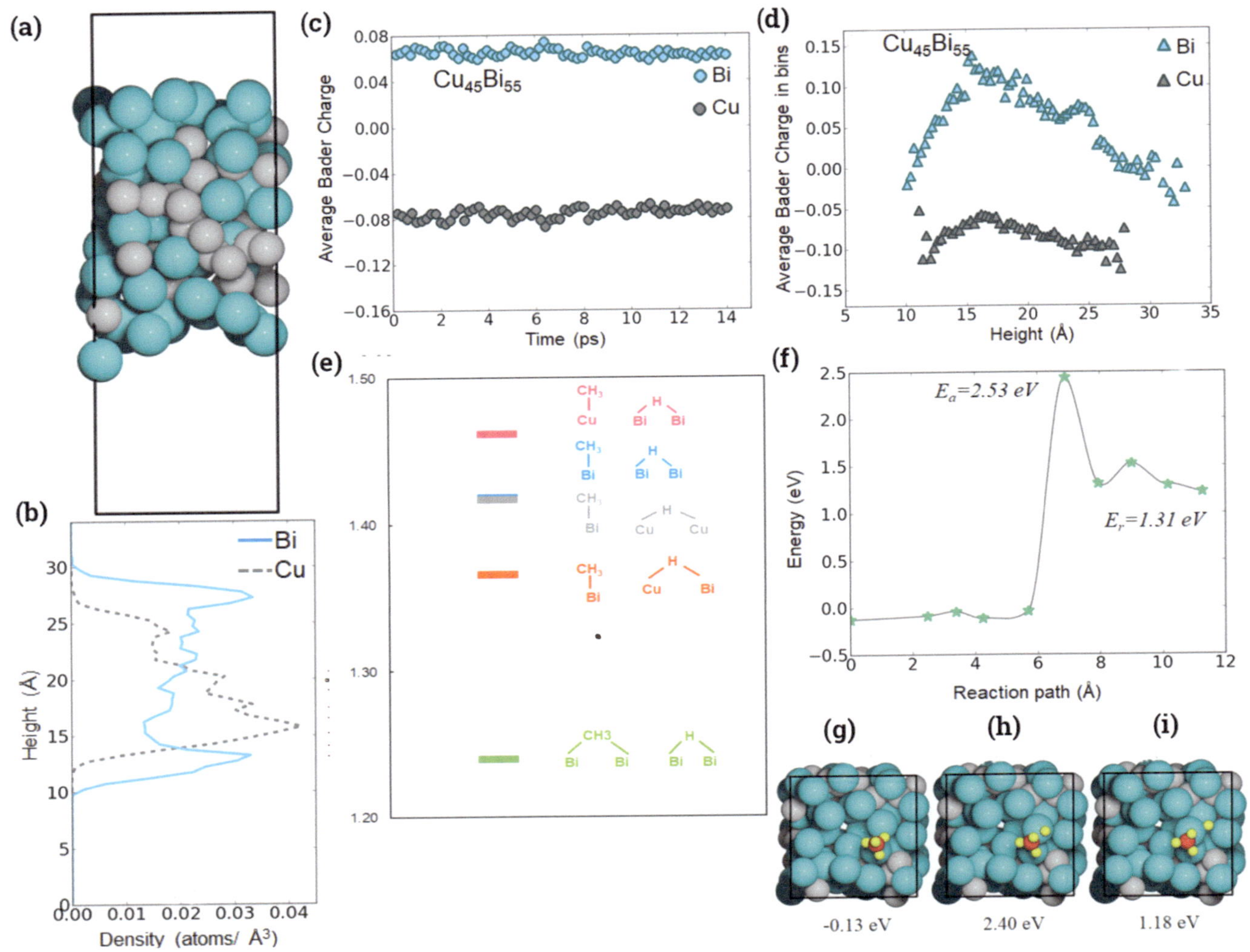

Fig. 5 **a** Snapshot of $Cu_{45}Bi_{55}$ melt from a long AIMD run. **b** Average number density plotted as a function of liquid height. **c** Average charges on Cu and Bi atoms. **d** Average charges on Cu and Bi atoms as a function of liquid height. **e** Binding of CH_3 and H on different sites on a liquid surface. **f** Potential energy surface of CH_4 dissociation with most stable dissociated configuration on a random snapshot of $Cu_{45}Bi_{55}$ liquid. **g–i** Top views of reactant, transition, and product state, respectively. Adapted from Palmer et al. [38]

In an exploratory study, Rajaprabu et al. [47] computationally studied the efficacy of molten and gas-phase Na for activating CH_4. Previously, molten sodium (Na) has been reported to activate N_2 and produce NH_3 in the presence of H_2, thereby acting as a catalyst for ammonia synthesis [24]. Since sodium has a tendency to lose its valence electron and acts as a reducing agent, Rajaprabu et al. [47] theoretically investigated the viability of molten Na as a catalyst for methane pyrolysis. Due to the high vapor pressure of Na, the catalytic influence of sodium vapors within the ascending bubble or in the space above the melt must be taken into consideration. The authors examined 24 nascent reaction pathways of CH_4 pyrolysis in gas phase and computed an effective activation energy for methane dissociation to be 296 kJ/mol. A microkinetic analysis of the reaction network followed by sensitivity analysis suggested that only three reactions, those involving the interaction of CH_4 with Na and H, determined the overall rate of methane conversion. On the other hand, on a molten Na surface, the authors found that the barriers for methane decomposition are significantly less. The study concluded that liquid sodium serves as an active catalyst for methane pyrolysis, mirroring findings similar to those observed for tellurium, where the catalyst in the liquid phase outperformed the vapor-phase catalyst for methane pyrolysis.

2.1 Molten Halides

Molten salts are ionic melts having dissociated ions. Due to the ionic character of the species and several attributes, such

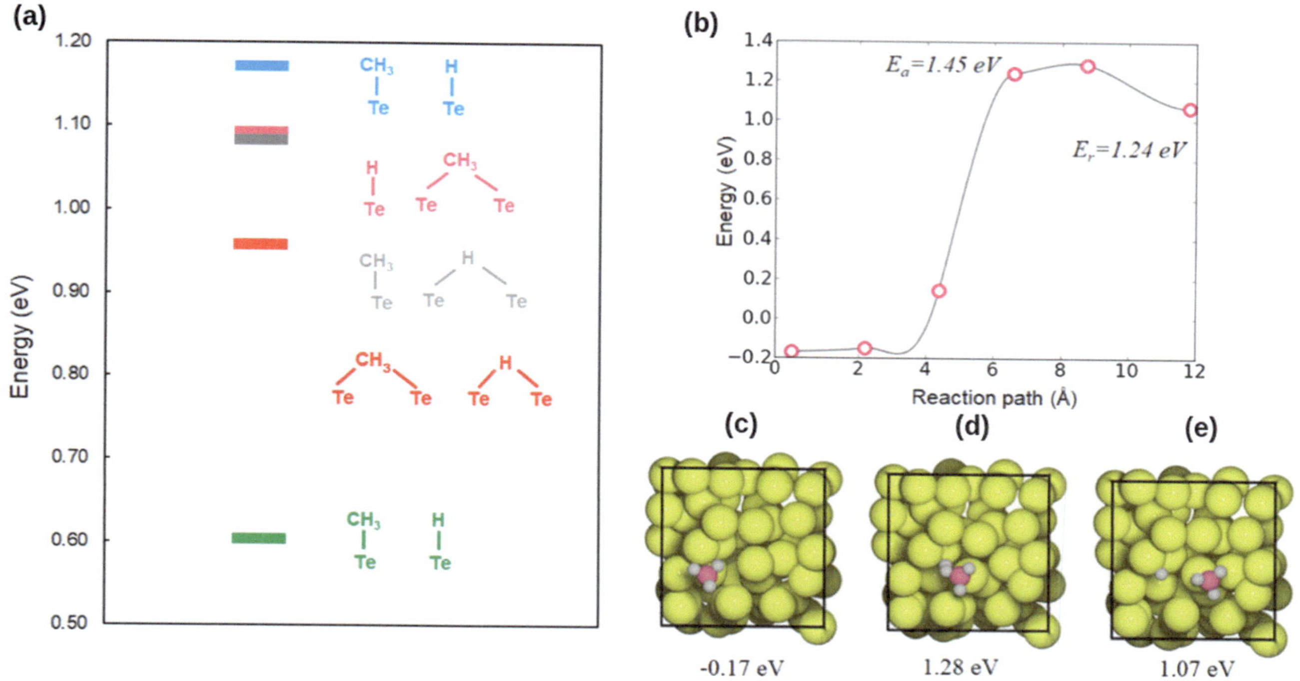

Fig. 6 **a** Binding of CH_3 and H on different sites on a liquid tellurium surface. **b** Potential energy surface of CH_4 dissociation with most stable dissociated configuration on a random snapshot of liquid tellurium. **c–e** Top views of reactant, transition, and product state, respectively. Adapted from Zeng et al. [67]

as high thermal stability, make them ideal media for many applications. Molten salt can engage in a reaction by serving as either a solvent or as a catalyst. A reaction in molten salts has several advantages over solid catalysts. One of them being effective dispersion of heat generated due to their high heat conductivity. Molten salts typically do not mix with the reaction products, which enables their effective separation. The dissociated ions create an electric field for polarizing reactants and may result in the formation of distinct products than solid catalysts. Furthermore, catalytically active sites within molten salts retain their activity and can be used, in principle, infinitely. Mixtures of molten salts containing transition metal atoms with multiple valence states display notable catalytic activity by creating complex ions upon dissolution. There are vast range of molten salts, with different cations and anions, considered in literature as a reaction media for carrying out several industrially important reactions [25, 59]. In this chapter, we focus on molten halides as catalyst for methane pyrolysis.

We begin by discussing simulations to understand methane pyrolysis in pure molten salts. Several monovalent and divalent metal halides have been tested in literature, which are summarized in Table 1. In a recent work, Palmer et al. [37] investigated the thermodynamics of stripping hydrogen off methane in four steps (see Fig. 7a) to generate carbon and hydrogen on several molten monovalent salts (NaCl, KCl, NaBr, and KBr). These calculations involved optimizing the locations of chemical species on the surface of a "glassy" slab of molten salt. The snapshot of the molten system was obtained at an arbitrary time from an equilibrated AIMD trajectory. The first step of breaking CH_3-H was found to be highly endergonic, ranging from 200 to 400 kJ/mol. The first step of methane dissociation is the rate-determining step, and therefore the activation barriers will also be high. The authors computed the transition state of methane dissociation on a slab on molten KCl (see Fig. 7b). Their calculations predicted an activation barrier of 317 kJ/mol, which matches qualitatively well with experimental findings.

KCl is poor catalyst for methane pyrolysis because the activation barriers for methane are very high. A recent experimental study [22] showed two surprising results. First, the activity of an eutectic melt of KCl/NaCl is enhanced dramatically by addition of only 1 wt% of $FeCl_3$. The apparent activation barriers reduce from ~300 to 178 kJ/mol. Second, even on increasing the concentration of $FeCl_3$ to 7 wt%, there was no considerable increase in activity. If Fe is the promoter, it is difficult to gage why activity will not increase with the increase in its concentration. Previously, the increase in activity was attributed to formation of complex ions. However, this behavior is different from doping with other metal halides, which also form complexes in KCl melt. For example, doping with small amounts of $MnCl_2$ and $AlCl_3$ does not bring about dramatic increase in activity of KCl toward methane pyrolysis.

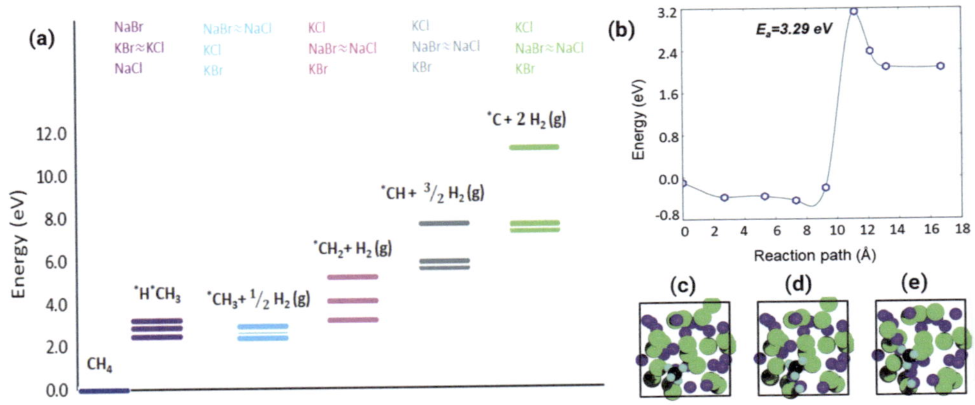

Fig. 7 **a** Free energy surface (at 1100 K) for methane dissociation on the surface of molten NaBr, NaCl, KCl, and KBr. **b** Potential energy surface for methane dissociation on molten KCl. **c–e** Top views of reactant, transition, and product state, respectively. Adapted from Palmer et al. [37]

Very recently, Dutta et al. [17] performed qualitatively accurate density functional theory-based molecular dynamics simulations, which were able to explain the remarkable behavior of $FeCl_3$ in KCl. The authors reported several interesting findings from this work. First, Fe forms several complexes in the melt of KCl (see Fig. 8a). The nature of the complex depends on the surrounding Cl ions and K ions. The charged state of Fe fluctuates between +3 and +2.

Second, in the +2 state, Fe ion withdraws an electron from several chlorine ions in the melt, which are not immediately bonded to Fe atom and are far away. These Cl ions act as centers for methane dissociation. Dutta et al. [17] extracted several snapshots of molten $FeCl_3$/KCl, when Fe was in +2 state, from long AIMD run, and performed transition-state calculations for methane dissociation on sites away from Fe. The activation energies lied in the range of 128 to 194 kJ/mol (see Fig. 8c), which shows a dramatic decrease in the activation energies.

Third, Dutta et al. [17] showed that $FeCl_3$ prefers to be in the bulk of the KCl system; the average energy in the bulk was lower than that on the surface (see Fig. 8b). Fourth, the authors tried forcing $FeCl_3$ on the surface and performing transition-state calculations on the Fe-Cl complex. However, they found that the reaction energies were greater than 200 kJ/mol, suggesting that Fe does not directly participate in the reaction. There is no reason to believe the behavior shown by Fe is specific and the activation pattern may be general, which is an appropriate topic for future exploration.

As the methane bubble rises in the molten reactor, it gets converted to hydrogen and carbon. Because of high vapor pressures of some of the metal halides at reaction temperatures (for example, $ZnCl_2$, $MnCl_2$), methane pyrolysis inside the bubbles cannot be overlooked. Additionally, in the head space above the melt there is a possibility of catalysis by vapor-phase metal halides. This raises number of interesting questions. First, will the metal halide catalyze the gas-phase CH_4 in the bubbles—i.e., is the apparent activation energy less than that of the uncatalyzed gas-phase CH_4 decomposition? Second, if the gas-phase metal halide acts as a catalyst what will be the nature of active sites? Third, what will be the reaction mechanisms and pathways? To answer these questions, Dutta et al. [16] performed quantum chemical calculations along with microkinetic modeling to explore the nascent pathways of methane decomposition in gas-phase $ZnCl_2$, based on the known fact that breaking of $CH_3{-}H$ bond is the rate-limiting step. The authors computed an effective activation energy for methane decomposition reaction of 227 kJ/mol (see Fig. 9a). The computed barrier was at par with several molten catalysts reported in literature, which implied that the gas-phase catalysis cannot be overlooked in the ascending bubbles. Dutta et al. [16] also performed sensitivity analysis based on degree of rate control to determine the rate-governing steps of the methane decomposition (see Fig. 9b) and found that overall rates were very sensitive to kinetic parameters of four reactions; among which direct reaction of metal halide with methane to form HCl was the most sensitive. The authors used this as the "descriptor" to explore the behavior of other gas-phase metal halides ($AlCl_3$, $FeCl_2$, $CuCl_2$, $NiCl_2$, and $CoCl_2$). Out of the several gas-phase metal halides investigated $CuCl_2$ was predicted to be the most reactive catalyst.

3 Future Perspectives

Several benefits of molten catalysts make them an ideal choice for catalysts in reactions where coking poses a significant challenge. In the envisageable "green" future, molten catalysts are, therefore, expected to play an important role as a reaction media for sustainable processes. Understanding the fundamental nature of catalytic sites and reaction bottlenecks can possibly aid in rational design of these molten catalyst. Over the past couple of decades, DFT simulations have established itself as a powerful tool to unravel the nature of catalytic sites and the underlying reaction mechanisms. With the advent of increased computational power and computationally tractable quantum mechanical simulations, it is now

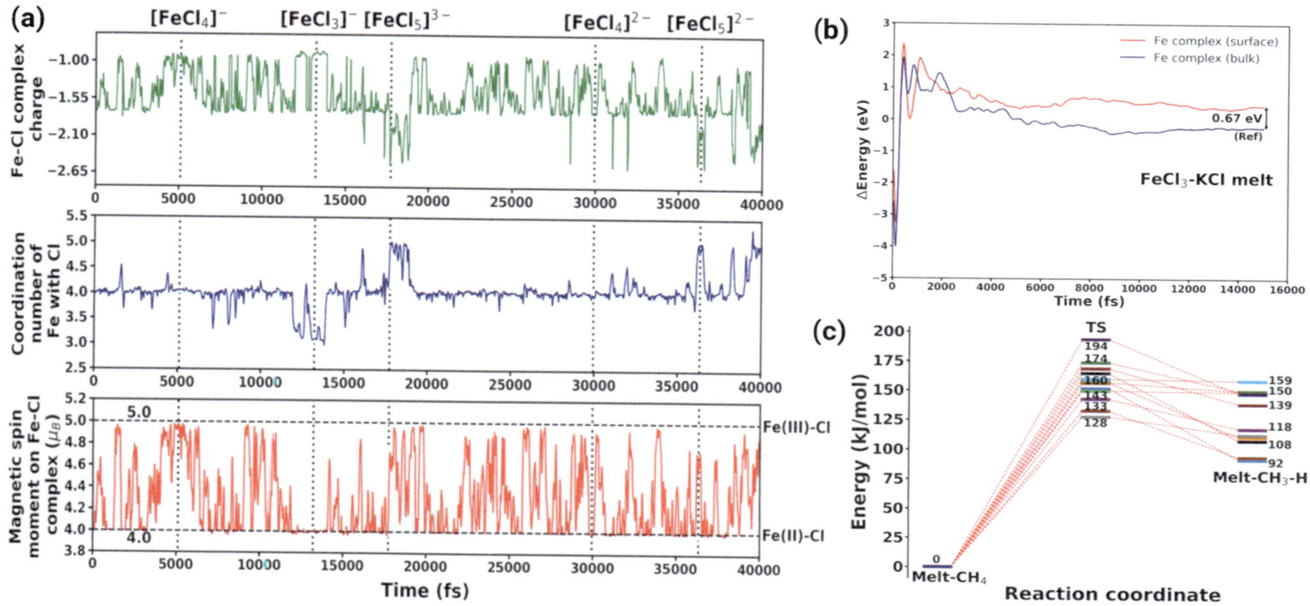

Fig. 8 **a** Top panel shows the Fe-Cl complex charge as a function of time in a AIMD run. Middle panel shows the coordination number of Fe with Cl atoms. And the bottom panel shows the magnetic moment of Fe-Cl complex in the melt. **b** Cumulative average energy of an equilibrated $FeCl_3$/KCl melt. **c** Potential energy surfaces of CH_4 dissociation for several snapshots of $FeCl_3$/KCl melt randomly taken from a long AIMD run. Adapted from Dutta et al. [17]

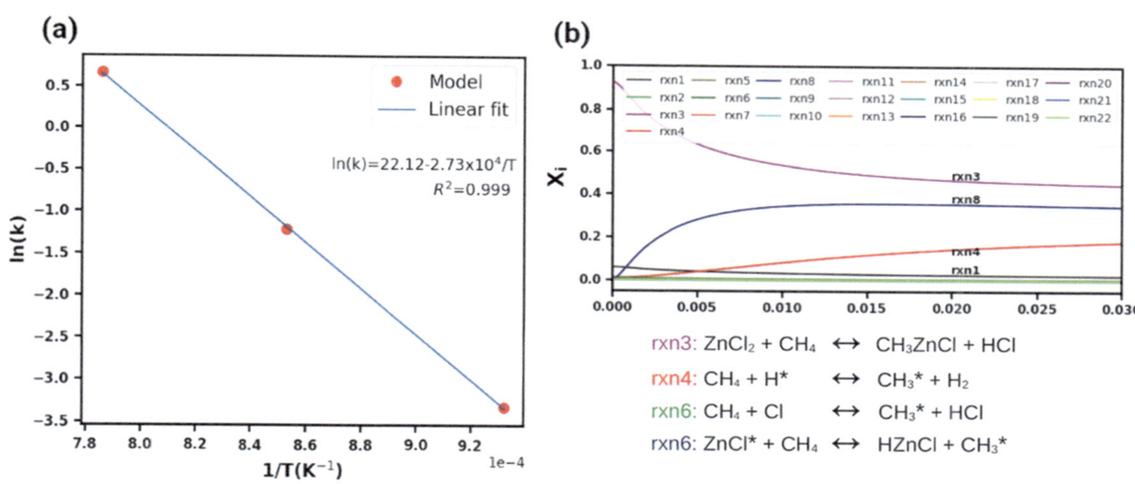

Fig. 9 **a** Arrhenius plot to extract activation energies for methane dissociation in gas-phase $ZnCl_2$. **b** Sensitivity analysis based on the degree of reaction control to identify rate-controlling steps. Adapted from Dutta et al. [16]

possible to perform qualitative simulations to understand several aspects of molten catalysts. In this chapter, we provide a perspective of first-principles simulations done to understand the nature of catalytic sites in molten catalysts for methane pyrolysis to hydrogen.

The area of molten catalysts is still young. Although the potential importance and economic feasibility has been established, commercial reactors have yet to be commissioned. The search for "better" molten catalyst for H_2 production from methane pyrolysis is an active area of research. First-principles calculations along with spectroscopic techniques can provide information on the molecular behavior and mechanisms of molten catalysts, which can expedite the research toward rational design of molten catalysts.

Acknowledgements N.S.R. and A.G. are thankful to MHRD for the financial support. V.A. thanks Prof. Horia Metiu and Dr. Sajal K. Dutta for stimulating discussions. V.A. acknowledges the financial support from Science and Engineering Board (SERB), grant no. CRG/2023/000623.

References

1. Abánades A, Rubbia C, Salmieri D (2013) Thermal cracking of methane into hydrogen for a CO_2-free utilization of natural gas. Int J Hydrogen Energy 38(20):8491–8496
2. Abánades A, Ruiz E, Ferruelo EM, Hernández F, Cabanillas A, Martínez-Val JM, Rubio JA, López C, Gavela R, Barrera G, et al (2011) Experimental analysis of direct thermal methane cracking. Int J Hydrogen Energy 36(20):12877–12886
3. Abdalla AM, Hossain S, Nisfindy OB, Azad AT, Dawood M, Azad AK (2018) Hydrogen production, storage, transportation and key challenges with applications: a review. Energy Convers Manag 165:602–627
4. Bae D, Kim Y, Ko EH, Han SJ, Lee JW, Kim M, Kang D (2023) Methane pyrolysis and carbon formation mechanisms in molten manganese chloride mixtures. Appl Energy 336:120810
5. Bartholomew CH, Farrauto RJ (2011) Fundamentals of industrial catalytic processes. Wiley
6. Capurso T, Stefanizzi M, Torresi M, Camporeale SM (2022) Perspective of the role of hydrogen in the 21st century energy transition. Energy Convers Manag 251:114898
7. Carapellucci R, Giordano L (2020) Steam, dry and autothermal methane reforming for hydrogen production: a thermodynamic equilibrium analysis. J Power Sources 469:228391
8. Chen G, Yu X, Ostrikov KK, Liu B, Harding J, Homm G, Guo H, Schunk SA, Zhou Y, Tu X, et al (2023) Methane up-carbonizing: a way towards clean hydrogen energy? Chem Eng J 146335
9. Chen L, Qi Z, Zhang S, Su J, Somorjai GA (2020) Catalytic hydrogen production from methane: a review on recent progress and prospect. Catalysts 10(8):858
10. Chen L, Song Z, Zhang S, Chang C-K, Chuang Y-C, Peng X, Dun C, Urban JJ, Guo J, Chen J-L, et al (2023) Ternary NiMo-Bi liquid alloy catalyst for efficient hydrogen production from methane pyrolysis. Science 381(6660):857–861
11. Cooper-Baldock Z, Perrelle TDL, Phelps C, Russell M, Ryan L, Schofield J, Nathan GJ, Jafarian M (2024) Preliminary evaluation of methods for continuous carbon removal from a molten catalyst bubbling methane pyrolysis reactor. Energies 17(2):290
12. Daeneke T, Khoshmanesh K, Mahmood N, De Castro IA, Esrafilzadeh D, Barrow SJ, Dickey MD, Kalantar-Zadeh K (2018) Liquid metals: fundamentals and applications in chemistry. Chem Soc Rev 47(11):4073–4111
13. Desai D, Karin T, Rivest JLB (2022) High throughput methane pyrolysis reactor for low-cost hydrogen production, September 27 2022. US Patent 11,453,584
14. Desai D, Rivest JLB (2020) System and method for pyrolysis using a liquid metal catalyst, December 1 2020. US Patent 10,851,307
15. Dipu AL (2021) Methane decomposition into CO_x-free hydrogen over a Ni-based catalyst: an overview. Int J Energy Res 45(7):9858–9877
16. Dutta SK, Ghosh S, Metiu H, Agarwal V (2022) Nascent decomposition pathways of CH_4 pyrolysis in gas-phase metal halides. J Phys Chem A 126(35):5900–5910
17. Dutta SK, Singh B, Metiu H, Agarwal V (2023) Increase of the catalytic activity of molten salts by doping: methane activation. J Phys Chem C 128(1):123–128
18. Erisman JW, Sutton MA, Galloway J, Klimont Z, Winiwarter W (2008) How a century of ammonia synthesis changed the world. Nat Geosci 1(10):636–639
19. Ertl G, Knözinger H, Weitkamp J, et al (2008) Handbook of heterogeneous catalysis, vol 1. Wiley-VCH Verlag GmbH& Co
20. Geißler T, Abánades A, Heinzel A, Mehravaran K, Müller G, Rathnam RK, Rubbia C, Salmieri D, Stoppel L, Stückrad S, et al (2016) Hydrogen production via methane pyrolysis in a liquid metal bubble column reactor with a packed bed. Chem Eng J 299:192–200
21. Geißler T, Plevan M, Abánades A, Heinzel A, Mehravaran K, Rathnam RK, Rubbia C, Salmieri D, Stoppel L, Stückrad S, et al (2015) Experimental investigation and thermo-chemical modeling of methane pyrolysis in a liquid metal bubble column reactor with a packed bed. Int J Hydrogen Energy 40(41):14134–14146
22. Kang D, Palmer C, Mannini D, Rahimi N, Gordon MJ, Metiu H, McFarland EW (2020) Catalytic methane pyrolysis in molten alkali chloride salts containing iron. ACS Catal 10(13):7032–7042
23. Kang D, Rahimi N, Gordon MJ, Metiu H, McFarland EW (2019) Catalytic methane pyrolysis in molten $MnCl_2$-KCl. Appl Catal B: Environ 254:659–666
24. Kawamura F, Taniguchi T (2017) Synthesis of ammonia using sodium melt. Sci Rep 7(1):11578
25. Kenney CN (1975) Molten salt catalysis of gas reactions. Catal Rev Sci Eng 11(1):197–224
26. Kim J, Oh C, Oh H, Lee Y, Seo H, Kim YK (2023) Catalytic methane pyrolysis for simultaneous production of hydrogen and graphitic carbon using a ceramic sparger in a molten NiSn alloy. Carbon 207:1–12
27. McConnachie M, Konarova M, Smart S (2023) Literature review of the catalytic pyrolysis of methane for hydrogen and carbon production. Int J Hydrogen Energy
28. McFarland EW, Upham DC, Zeng J, Palmer C, Su S, Mannini D, Kang D, Rahimi N, Metiu H, Gordon MJ (2023) Simultaneous reaction and separation of chemicals, November 14 2023. US Patent 11,814,285
29. Megía PJ, Vizcaíno AJ, Calles JA, Carrero A (2021) Hydrogen production technologies: from fossil fuels toward renewable sources. A mini review. Energy Fuels 35(20):16403–16415
30. Msheik M, Rodat S, Abanades S (2021) Methane cracking for hydrogen production: a review of catalytic and molten media pyrolysis. Energies 14(11):3107
31. Nishikawa Y, Ohtsuka Y, Ogihara H, Rattanawan R, Gao M, Nakayama A, Hasegawa J, Yamanaka I (2020) Catalytic mechanism of liquid-metal indium for direct dehydrogenative conversion of methane to higher hydrocarbons. ACS Omega 5(43):28158–28167
32. Niu J, Wang Y, Qi Y, Dam AH, Wang H, Zhu Y-A, Holmen A, Ran J, Chen D (2020) New mechanism insights into methane steam reforming on Pt/Ni from DFT and experimental kinetic study. Fuel 266:117143
33. Oblad AG, Milliken TH Jr, Boedeker ER (1956) Production of hydrogen and carbon, August 28 1956. US Patent 2,760,847
34. Ogino Y (1981) Catalysis by molten metals and molten alloys. Catal Rev Sci Eng 23(4):505–551
35. Ogino Y (2020) Catalysis and surface properties of liquid metals and alloys. CRC Press
36. Oni AO, Anaya K, Giwa T, Di Lullo G, Kumar A (2022) Comparative assessment of blue hydrogen from steam methane reforming, autothermal reforming, and natural gas decomposition technologies for natural gas-producing regions. Energy Convers Manag 254:115245
37. Palmer C, Tarazkar M, Gordon MJ, Metiu H, McFarland EW (2021) Methane pyrolysis in low-cost, alkali-halide molten salts at high temperatures. Sustain Energy Fuels 5(23):6107–6123
38. Palmer C, Tarazkar M, Kristoffersen HH, Gelinas J, Gordon MJ, McFarland EW, Metiu H (2019) Methane pyrolysis with a molten Cu–Bi alloy catalyst. ACS Catal 9(9):8337–8345
39. Palmer C, Upham DC, Smart S, Gordon MJ, Metiu H, McFarland EW (2020) Dry reforming of methane catalysed by molten metal alloys. Nat Catal 3(1):83–89

40. Parkinson B, Matthews JW, McConnaughy TB, Upham DC, McFarland EW (2017) Techno-economic analysis of methane pyrolysis in molten metals: decarbonizing natural gas. Chem Eng Technol 40(6):1022–1030
41. Parkinson B, Patzschke CF, Nikolis D, Raman S, Dankworth DC, Hellgardt K (2021) Methane pyrolysis in monovalent alkali halide salts: kinetics and pyrolytic carbon properties. Int J Hydrogen Energy 46(9):6225–6238
42. Parra AAM, Agar DW (2017) Molten metal capillary reactor for the high-temperature pyrolysis of methane. Int J Hydrogen Energy 42(19):13641–13648
43. Patzschke CF, Parkinson B, Willis JJ, Nandi P, Love AM, Raman S, Hellgardt K (2021) Co-Mn catalysts for H_2 production via methane pyrolysis in molten salts. Chem Eng J 414:128730
44. Paxman D, Trottier S, Nikoo M, Secanell M, Ordorica-Garcia GJEP (2014) Initial experimental and theoretical investigation of solar molten media methane cracking for hydrogen production. Energy Procedia 49:2027–2036
45. Pérez BJL, Jiménez JAM, Bhardwaj R, Goetheer E, van Sint Annaland M, Gallucci F (2021) Methane pyrolysis in a molten gallium bubble column reactor for sustainable hydrogen production: proof of concept & techno-economic assessment. Int J Hydrogen Energy 46(7):4917–4935
46. Plevan M, Geißler T, Abánades A, Mehravaran K, Rathnam RK, Rubbia C, Salmieri D, Stoppel L, Stückrad S, Wetzel T (2015) Thermal cracking of methane in a liquid metal bubble column reactor: experiments and kinetic analysis. Int J Hydrogen Energy 40(25):8020–8033
47. Surya Rajaprabu N, Bhagat SK, Metiu H, Agarwal V (2024) Activation of methane in vapor and molten sodium catalyst. Accepted in JPCC
48. Ross JRH (2011) Heterogeneous catalysis: fundamentals and applications. Elsevier
49. Rupprechter G (2017) Popping up to the surface. Nat Chem 9(9):833–834
50. Schultz I, Agar DW (2015) Decarbonisation of fossil energy via methane pyrolysis using two reactor concepts: fluid wall flow reactor and molten metal capillary reactor. Int J Hydrogen Energy 40(35):11422–11427
51. Serban M, Lewis MA, Marshall CL, Doctor RD (2003) Hydrogen production by direct contact pyrolysis of natural gas. Energy Fuels 17(3):705–713
52. Shaner MR, Atwater HA, Lewis NS, McFarland EW (2016) A comparative technoeconomic analysis of renewable hydrogen production using solar energy. Energy Environ Sci 9(7):2354–2371
53. Sheil A, Konarova M, Smart S (2023) Catalytic alkali and transition metal cations to produce low-emission hydrogen from methane pyrolysis. ACS Sustain Chem Eng 11(30):11248–11258
54. Skinner GB, Ruehrwein RA (1959) Shock tube studies on the pyrolysis and oxidation of methane. J Phys Chem 63(10):1736–1742
55. Sorcar S, Rosen BA (2023) Methane pyrolysis using a multiphase molten metal reactor. ACS Catal 13(15):10161–10166
56. Steinberg M (1999) Fossil fuel decarbonization technology for mitigating global warming. Int J Hydrogen Energy 24(8):771–777
57. Steinberg M, Dong Y (1998) Method for converting natural gas and carbon dioxide to methanol and reducing CO_2 emissions, June 16 1998. US Patent 5,767,165
58. Stiegel GJ, Ramezan M (2006) Hydrogen from coal gasification: an economical pathway to a sustainable energy future. Int J Coal Geol 65(3-4):173–190
59. Sundermeyer W (1965) Fused salts and their use as reaction media. Angewandte Chemie International Edition in English 4(3):222–238
60. Taccardi N, Grabau M, Debuschewitz J, Distaso M, Brandl M, Hock R, Maier F, Papp C, Erhard J, Neiss C et al (2017) Gallium-rich Pd-Ga phases as supported liquid metal catalysts. Nat Chem 9(9):862–867
61. Tang Y, Peng P, Wang S, Liu Z, Zu X, Yu Q (2017) Continuous production of graphite nanosheets by bubbling chemical vapor deposition using molten copper. Chem Mater 29(19):8404–8411
62. Tarazkar M, Metiu H, Kristoffersen HH (2021) Properties of methane and carbon adsorbed at the interface between molten NaBr and Ni (111). J Phys Chem C 125(7):3980–3987
63. Tyrer D (1931) Production of Hydrogen, US Patent 1,803,221, Apr. 28, 1931
64. Upham DC, Agarwal V, Khechfe A, Snodgrass ZR, Gordon MJ, Metiu H, McFarland EW (2017) Catalytic molten metals for the direct conversion of methane to hydrogen and separable carbon. Science 358(6365):917–921
65. Wang K, Li WS, Zhou XP (2008) Hydrogen generation by direct decomposition of hydrocarbons over molten magnesium. J Mol Catal A: Chem 283(1–2):153–157
66. Zaghloul N, Kodama S, Sekiguchi H (2021) Hydrogen production by methane pyrolysis in a molten-metal bubble column. Chem Eng Technol 44(11):1986–1993
67. Zeng J, Tarazkar M, Pennebaker T, Gordon MJ, Metiu H, McFarland EW (2020) Catalytic methane pyrolysis with liquid and vapor phase tellurium. ACS Catal 10(15):8223–8230
68. Zhi X, Xie F, Zavabeti A, Li GK, Harvie DJE, Kashi E, Batterham RJ, Liu JZ (2023) Fundamentals, applications, and perspectives of liquid metals in catalysis: an overview of molecular simulations. Energy Fuels 37(23):17875–17891
69. Zuraiqi K, Zavabeti A, Allioux F-M, Tang J, Nguyen CK, Tafazolymotie P, Mayyas M, Ramarao AV, Spencer M, Shah K, et al (2020) Liquid metals in catalysis for energy applications. Joule 4(11):2290–2321

Application of Engineered Biochar for CO_2 Capture

Mohit Kumar, Himanshu Sharma, Raju Kumar Gupta, and Anand Singh

1 Introduction

In the face of escalating concerns about climate change and the urgent need for sustainable solutions, the exploration of innovative technologies for carbon dioxide (CO_2) capture has become imperative. Fossil fuels are extensively consumed for economic growth and increased human activity, and as a result, the amount of CO_2 emitted into the atmosphere has been continuously increasing [1]. Currently, the concentration of CO_2 in the atmosphere is reported to be over 400 parts per million (ppm) with CO_2 being the predominant (contributing to 90%) anthropogenic greenhouse gas on a global scale [2]. More than 80% of CO_2 in the environment today [3] originates from the burning of fossil fuels, such as coalmine solid fuel, petroleum oil, and natural gas industries [4]. This amounts to around 32 Gt CO_2 generated annually [5]. It is anticipated that this figure will triple by 2050 due to the ongoing increase in the level of CO_2 [6]. If CO_2 concentration continues to rise, climate change will have a significant and long-lasting impact on ecosystems and humans. According to the projections from IPCC, it is anticipated that there will be a significant increase in the average global temperature of approximately 1.9 °C by the year 2100, corresponding to a CO_2 concentration of 570 ppm [7].

It is hard to cease CO_2 emissions from sources of emission right away due to society's expanding demand for energy production. However, efforts are ongoing to curtail CO_2 emissions and prevent global warming from getting worse. In December 2015, 195 nations agreed in Paris to regulate greenhouse gas emissions and ensure that the global average temperature remains under 2 °C above pre-industrial levels. However, a significant reduction in CO_2 emissions is required to meet this goal.

The entire CO_2 emissions can be generally analyzed in three different scopes. Scope 1 or direction emission is from the combustion of fuels. It is generally counted within the plant/factory's boundaries only. Scope 2, or indirect emissions, are associated with the purchase of utilities and typically originate from sources outside the facility's boundaries. Scope 3 or direct emissions relates to the whole value chain such as equipment procurement, building workings, raw material obtaining, and product dispatch.

Indian industrial and power sector contributed approximately 1600 million metric tons per annum (mtpa) of CO_2 emission (approx. 60%) of the total 2,600 mtpa in 2020. The remaining 40% of CO_2 emissions come from dispersed sources, including farming, transport, and constructions [8]. This number is expected to rise to 2300 mtpa by the year 2030 due to a continued increase in the energy demand in industries and continued urbanization [8].

1.1 Industrial CO_2 Emission Sources in India

The major industrial sectors responsible for generating CO_2 are depicted in Fig. 1. With its position as the world's second largest user and third largest producer of electricity, India's power industry is a significant player in the global energy landscape. On a global scale, the power sector is the leading contributor to greenhouse gas (GHG) emissions, comprising 42% of the world's total anthropogenic CO_2 emissions in 2019, as reported by the International Energy Agency (IEA). In India, 51% of the country's total human-made emissions are attributed to the power sector, as reported by the IEA [8].

M. Kumar · H. Sharma (✉) · R. K. Gupta · A. Singh
Department of Sustainable Energy Engineering, IIT Kanpur, Kanpur, India
e-mail: sharmah@iitk.ac.in

H. Sharma · R. K. Gupta
Department of Chemical Engineering, IIT Kanpur, Kanpur, India

A. Singh
Department of Chemistry, IIT Kanpur, Kanpur, India

S. Narayanan et al. (eds.), *Materials, Devices and Systems for Sustainability*, IITK Directions 8,
https://doi.org/10.1007/978-981-96-7295-0_7

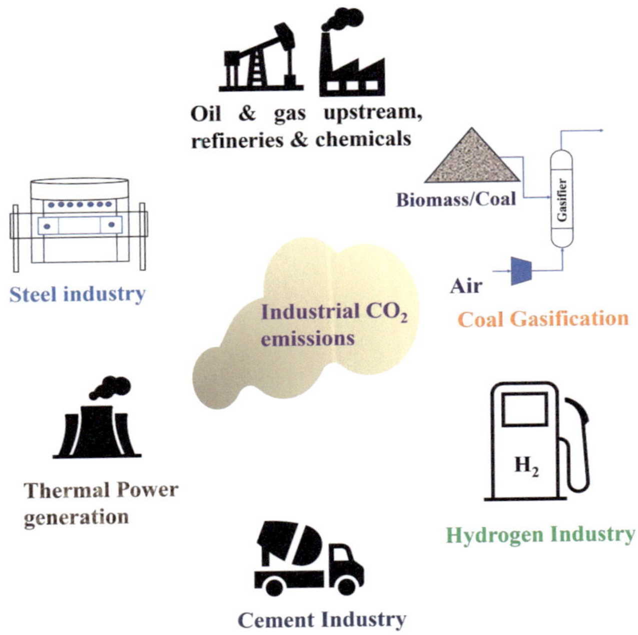

Fig. 1 Major industrial CO_2 emissions sources

Approximately 10% of India's total CO_2 emission comes from the energy-intensive iron and steel sector, which is mostly focused on processing virgin resources such as coal and iron ore. CO_2 emission occurs from three different routes in the steel industry: (i) Basic Oxygen Furnaces (BOF), ii) Electric Arc Furnace (EAF), and (iii) Induction Furnace (IF) of the steel and iron-making industry.

As per emission data from various industrial practices and production information for the fiscal year 2019–20, the aggregate CO_2 emissions from the three pathways are approximately 240 mtpa. With the current trajectory of steel demand, it's plausible to anticipate that the making of steel could rise to 190–200 mtpa by 2030, with demand increasing at a CAGR of 6%. According to steel, demand is growing, and this trend would lead to a rise in CO_2 emissions. According to estimates based on the corresponding CO_2 emission intensities, India's steel sector will produce 450 million metric tons of CO_2 emissions annually by 2030 [8].

India ranks as the world's second largest cement producer following China. Cement production has witnessed significant growth over the past 5 years, increasing from 274 mtpa in the fiscal year 2015–2016 to 381 mtpa in the fiscal year 2021–2022, reflecting a CAGR of 6% [8]. In 2018, cement manufacturing accounted for approximately 12% or 264 mtpa of humans caused CO_2 releases in the nation, according to the IEA. With the continual elevation in cement requirement and manufacturing in the foreseeable future, these emissions are expected to increase. CO_2 emissions exhibit significant variation from one plant to another, influenced by factors including type of product, operational efficiency, fuel choices, and production capacity, among others [8].

Another energy-intensive sector that generates materials for numerous downstream industries is the chemical industry. Indian chemical industry is placed third in Asia and sixth globally, respectively. One of the biggest centers for chemical production is India, and over the next 5–10 years, the sector is predicted to grow rapidly, increasing CO_2 emissions dramatically [8]. Roughly 160 million tons of CO_2 emissions (or 7% of total emissions) came from the chemical sector in 2020. Growth in fertilizers, petrochemicals, and specialty chemicals is predicted to drive this increase to roughly 220 million tons by 2030.

Coal remains the predominant and most vital conventional solid fuel in India. Synthesis gas is produced when coal undergoes gasification. This gas can be used to make various products. Some of these products include Dimethyl Ether (DME), methanol, ethanol, hydrogen, olefins, and ammonia. During gas processing, CO_2 gas is generated as a byproduct. Several processes contribute to this, such as gasification and the water gas shift reaction. Other steps like acid gas removal also produce CO_2.

1.2 CO_2 Capture Techniques

CO_2 capture methods such as absorption and adsorption have been studied in past years. Using amine for carbon capture is a mature technique and is used extensively in the industry [8]. However, the corrosive nature of amines reduces the working life of the equipment and therefore increases the cost of the process. Moreover, the absorbent is also reactive with other acid gases in the gas stream subsequently decreasing the activity of the CO_2 being absorbed. The adsorption process that involves solid–gas interaction utilizes solid adsorbent for the separation of CO_2 from the gas stream. Solid adsorbents like amine-supported silica, carbonaceous materials, zeolites, and metal–organic frameworks (MOFs) have become increasingly favored over liquid absorbents because of their superior adsorption capacity and lower cost. However, because of some of their disadvantages, the use of another solid adsorbing materials, including metal–organic frameworks and zeolites, is restricted. The structure of MOFs may potentially be destroyed by the gas if it includes contaminants. Although zeolites are more robust and ideal for gas adsorption, their adsorption ability diminishes with increasing temperature. Out of various techniques for the reduction of CO_2 emissions, capturing CO_2 using different types of solid adsorbents is one of the most suitable from the source before getting mixed into the environment. Two key factors to scale up CO_2 capture technology are its high adsorption capacity and the cost-effectiveness of the process [1].

1.3 Biochar

Biochar is a kind of substance that is more carbon-rich, strongly aromatized, insoluble, and stable than the solid adsorbents mentioned above [9]. Biochar is a carbon-rich material. It has a fine-grained and porous structure. It is created by heating the biomass. This process occurs in an environment with limited oxygen content. The thermal decomposition of biomass results in biochar [10]. It originates from using thermochemical conversion of waste within an environment where oxygen is limited. Biochar is produced from biomass, including agricultural waste such as wheat straw, groundnut shells, wood waste, agro-industrial wastes, and municipal solid sludge. Large volumes of biomass are generated every year from different sectors and are presented in Fig. 2.

The major elemental compositions such as Nitrogen (N), Carbon (C), Oxygen (O), and Hydrogen of biochar vary depending upon the experimental condition and feedstocks. In general, the carbon content in biochar is more than 40% of the total elemental composition [1, 11]. Biochar is produced from different types of raw feedstocks using the pyrolysis technique. The biochar shows various physicochemical properties such as surface functional group, elemental constituents, pore size and structure, specific surface areas, and surface oxygen. According to the properties of the biochar, it can be used in various applications such as solid adsorbents, soil modifications, catalysis, and probe materials. Biochar has a desired surface structure that makes it a viable substitute for traditional solid carbon-based catalytic materials, which are neither sustainable nor economically friendly. Biochar may be utilized as an efficient adsorbent for CO_2 capture. The interdisciplinary nature of this topic brings together expertise from various fields such as materials science, environmental engineering, and bioenergy research. This book chapter aims to provide a comprehensive resource for researchers, academics, and industry professionals interested in understanding the potential and challenges associated with the application of engineered biochar for CO_2 capture. By examining the latest advancements and emerging trends in this rapidly evolving field, we hope to contribute to the ongoing dialogue surrounding sustainable practices and climate change mitigation.

Energy Crops: Plant exclusively grown their wood from marginal lands for energy generation e.g. prosopis, bamboo

Agro-Industrial Waste: Wastes from paper mills, molasses from sugar refineries, pulp wastes, textile fiber waste

Agricultural Wastes: Straws of cereals and pulses, seed coats of oilseed, crop wastes, sugarcane trash, rice husk, coconut shell

Municipal Solid Waste: Biodegradables such as food and kitchen waste, green waste, paper waste like fabric clothes

Forest Waste: Logs, chips, barks, leaves, forest industry-based product like sawdust

Fig. 2 Different types of biomasses for biochar production

2 Current Developments in CO_2 Capture

During CO_2 capture, CO_2 is captured on or after gas streams that are emitted from manufacturing activities including power production plants, gasification plants, chemical manufacture, cement production, and steel production. Technologies for capturing CO_2 fall into the following three main categories: post-combustion, pre-combustion, and oxy-fuel combustion capture.

(I) **Post-combustion technology**

Post-combustion CO_2 capture involves capturing CO_2 produced due to the burning of fossil fuels. Burning fossil fuels such as petroleum, natural gas, and coal results in flue gas that has high nitrogen content, and the concentration of CO_2 is about 3–15% by volume. Post-combustion CO_2 capture technology is applied downstream of the combustion process. It typically involves the use of liquid solvents or solid adsorbents to selectively absorb CO_2 from the flue gas (Fig. 3). Note that the pressure of the exhaust stream is low in post-combustion CO_2 capture. Post-combustion systems are not cost-effective and energy-intensive because chemical sorbents are needed for CO_2 capture due to the low partial pressure of CO_2, energy-intensive regeneration of sorbents, and the corrosive nature of amine-based absorbents [8].

(II) **Pre-combustion technology**

This entails treating petroleum and natural gas upstream before they are burned to remove CO_2. In gasification, fossil fuels undergo a transformation. They are converted into a mixture. This mixture consists of hydrogen and carbon monoxide. The resulting substance is called syngas. Syngas is created before combustion occurs. CO_2 is separated from this syngas stream before it is burned, often using methods such as physical absorption and pressure swing adsorption (Fig. 3). The primary distinction among pre-burning and post-burning lies in favoring the earlier when the gas stream

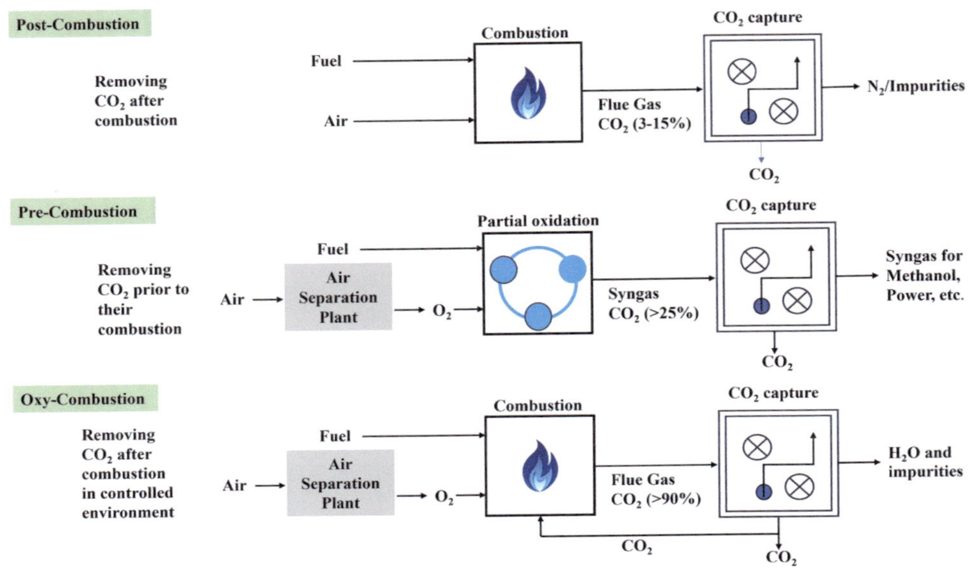

Fig. 3 Diagram illustrating Post-, Pre-, and Oxy-fuel combustion processes [8]

exhibits a higher concentration or partial pressure of CO_2, as seen in fossil fuel gasification, natural gas-based hydrogen (H_2) production, or sour gas processing. This is the main distinction between the two processes. The solvent regeneration process, when using physical solvents, requires no breaking of chemical bonds, so there is a significantly lower regeneration cost.

(III) **Oxy-fuel combustion technology**

Oxy-fuel combustion is a process that utilizes an oxygen-enriched environment instead of regular air for burning fossil fuels. This approach produces a flue gas primarily consisting of carbon dioxide (CO_2) and water vapor. CO_2 can be separated after vapor of H_2O with ease. After separation, the remaining CO_2 stream becomes concentrated. This concentrated CO_2 can then be captured (Fig. 3). This technology applies to both power plants and industrial processes. Unlike pre-burning and post-burning carbon capture machineries, which are already in commercial use, oxy-fuel burning is still under maturity. As a promising innovation, this technology aims to enable near-zero emissions by minimizing the production of other pollutants and maximizing the capture of CO_2 [8]. Though in its early stages, oxy-fuel combustion holds significant potential for supporting global carbon reduction goals, particularly in energy-intensive sectors.

2.1 Direct Air Capture (DAC)

Since it is not reliant on the origin or intensity of the emission stream, DAC is a method of carbon collection that may be applied widely. It directly captures CO_2 (at 415 ppm) from the air. DAC is yet in its infancy, hence the financial viability and scope of its processes have not yet been determined.

3 Production Strategies of Biochar

Biochar production is the process of transforming biomass, such as agricultural residues or organic waste, into a stable, carbon-rich material through a thermal treatment technique known as pyrolysis. This process involves heating the organic matter under controlled conditions with little to no oxygen, which prevents complete combustion and facilitates the breakdown of complex compounds. Pyrolysis is a versatile technology that not only produces biochar but also generates valuable byproducts like bio-oil and syngas, depending on the method used. There are various production strategies for biochar, each with its advantages and applications. Here are some common biochar production methods:

3.1 Traditional Kilns

Traditional kilns, such as pit or mound kilns, have been historically used for biochar production. These are simple structures where biomass is stacked, covered, and then ignited. The limited oxygen availability in the covered environment promotes pyrolysis. While these methods are straightforward, they may not provide precise control over the pyrolysis conditions.

Table 1 Operating condition and different products during thermal conversion of biomasses [1]

Thermochemical system	Reaction Temp (°C)	Raping rate (°C/min)	Holding time	Feedstock size (mm)	Biochar output (%)	Primary products
Slow rate pyrolysis	400–500	5–7	> 1 h	5–50	25–35	Mainly solid biochar
Fast rate pyrolysis	850–1250	6000–12,000	Under 2 s	Less than1	10–25	Primarily bio-oil
Intermediary pyrolysis	400–650	60–600	10–30 s	1–5	25–40	Mix of liquid, solid, and gas
Flashy pyrolysis	Above 1000	Over 60,000	Less than 1 s	Below 0.5	10–25	Primarily bio-oil
Gasification	Over 500	1–50	10–20 s	0.2–10	5–10	Produces primarily syngas

3.2 Improved Combustion Kilns:

Modern kilns, such as retort kilns and flame cap kilns, offer better control over the pyrolysis process. Retort kilns, for example, allow for better management of temperature and oxygen levels during pyrolysis, leading to more efficient biochar production.

3.3 Continuous Pyrolysis Systems

Continuous pyrolysis systems involve a continuous flow of biomass through a reactor where pyrolysis occurs. These systems are designed to optimize the heating and cooling processes, increasing the efficiency of biochar production. Continuous pyrolysis can be achieved through various reactor designs, such as screw conveyor pyrolyzers and fluidized bed reactors.

3.4 Microwave Pyrolysis

Microwave-assisted pyrolysis is a relatively novel method that uses microwave energy to heat biomass and facilitate pyrolysis. This method can provide rapid and controlled heating, leading to efficient biochar production. It has the potential for smaller-scale operations and decentralized applications.

3.5 Hydrothermal Carbonization (HTC)

HTC contains the transformation of wet material, such as organic waste or sludge, into biochar through a combination of heat and pressure in a water-rich environment. This process can produce hydrochar, which has different properties than traditional biochar.

3.6 Gasification

Gasification is another method where biomass is converted into syngas (a mixture of carbon monoxide and hydrogen) and biochar through partial combustion. The syngas can be used for energy production, and the biochar is a byproduct (Table 1).

3.7 Batch Pyrolysis

It is a thermal treatment route that produces char, pyrolytic oil, and gaseous combination of carbon dioxide, hydrogen, and light hydrocarbons using thermal degradation of organic waste biomasses in an oxygen-free atmosphere. Generally, the main constituents of biomass are cellulose, hemicellulose, and lignin. During the thermal degradation process, the primary constituents of biomass decompose with their set of various reactions at their corresponding temperatures such as cross-linked bonds, the breakdown of polymers, and destruction. The yields of the solid product from pyrolysis depend on several factors. These factors include the physical quality of the biomass feed. The temperature after the reaction also plays a role. The rate of heating affects the yields as well. Additionally, the reaction time influences the distribution of carbon in the solid product. The pyrolysis process is generally divided into slow pyrolysis, intermediate pyrolysis, fast pyrolysis, and flash pyrolysis. These are categorized based on the heating rate. The residence time of the vapors inside the reactor is also a distinguishing factor.

3.7.1 Slow Pyrolysis

Slow pyrolysis is a widely used method. It converts various raw materials into solid biochar. It also produces non-condensable gases and bio-oils. The feedstocks undergo thermal degradation via gradual pyrolysis in a neutral environment at temperatures between 300 and 800 °C [12]. Slow pyrolysis uses lower ramp rates. These rates range from 5 to 10 °C/min. It also requires longer vapor residence times. These times can range from a few minutes to several hours [13]. As the heating temperature and rate rise, the yield of biochar falls [8]. Conversely, increasing the pyrolysis temperature leads to higher carbon content and specific surface area, potentially enhancing CO2 capture capacity. Slower heating rates encourage polymerization reactions. These reactions occur during the lower-temperature stages of pyrolysis. They help convert aromatic compounds into bio-oil. This conversion results in the formation of coke.

To study their impact on the performance of pyrolytic products, a variety of carrier gases, both oxidative and reductive, have been used in the pyrolysis process. These gases include CO2, steam, and CO. They also contain CH_4 and H_2. In some cases, these gases are found in combinations [14]. It has been shown that char morphology gets more established and disordered (with more carbon active sites) in an impure pyrolysis environment that includes H_2O, which results in a larger char reactivity [14]. When pine sawdust was pyrolyzed in H_2 between 400 and 600 °C, the amount of char produced was lower than in an inert (Ar) environment. This suggests that H_2 can help release vapors. It does so by interacting with carbon at higher temperatures. This interaction reduces the amount of char produced [15]. Therefore, the production and characteristics of biochar, bio-oil, and non-condensable gases including specific surface area, porosity, ultimate analysis, and functional group of chemical compounds that are required for various applications may be changed using CO_2. Thus, valuable products may be produced by using CO_2 slow pyrolysis to make tailored biochar.

3.7.2 Intermediate Pyrolysis

The process condition for intermediate pyrolysis lies between slow and fast pyrolysis. Pyrolysis is usually executed at temperatures extending from 300 and 600 °C and with a vapor residence period of 2–4 s [16]. Earlier studies state that the benefits of intermediate pyrolysis include low operating temperature, processing a wide variety of raw materials, and a superior final output [17]. Biochar and bio-oil may be simultaneously produced using intermediate pyrolysis, which has an excellent product distribution. Prior research has demonstrated that biochar produced during intermediate pyrolysis is more suited for adsorption processes in hydrometallurgy [18]. Furthermore, waste that cannot be handled by rapid pyrolysis, such as those with a significant amount of water, low heat value, and high ash content, can be efficiently treated by intermediate pyrolysis. Both the need for recycling waste and the need for biochar are rising now. Therefore, converting solid waste into biochar by intermediate pyrolysis looks like a feasible alternative.

3.7.3 Fast Pyrolysis

Fast pyrolysis is a thermochemical process of conversion that produces liquid bio-oil and trace quantities of solid byproducts in the absence of oxygen from trash or underdone biomass material such as food crop residues, agricultural waste, forestry leftovers, and community trash [19]. Typical considerations for fast pyrolysis include an extreme heat-up speed exceeding 100 °C/min, coupled with a brief reaction time of feedstock material fragments and warm vapor (ranging from 0.5 to 2 s) at temperatures between 400 and 600 °C. In the fast pyrolysis process, a bio-oil product can be produced. This bio-oil makes up 70–75 weight% of the initial dried biomass. The exact amount depends on several factors. These include the qualities of the unprocessed material, the pyrolysis temperature, and the vapor residence time [20]. Shorter particle sizes (1–2 mm) in the raw materials are needed for fast pyrolysis to get high efficiency in transferring heat from the source of heating to the biomass components. Furthermore, rapid pyrolysis necessitates the use of a device that can swiftly remove steam to regulate a short residence period. The proportion of pyrolysis products will also vary based on operation environments. The solid char yield can range from 5 to 25%, while the gas product yield can range from 10 to 20%. The major product in fast pyrolysis is the liquid bio-oil [21].

3.7.4 Flash Pyrolysis

Fast pyrolysis and flash pyrolysis are comparable, however, flash pyrolysis will heat something significantly more quickly. Flash pyrolysis is characterized by high reaction temperatures (900–1300 °C) and fast heating rates (>1000 °C/s). The naturally occurring raw materials have tiny particles ranging from 105 to 250 μm [22]. Fast pyrolysis produces more liquid pyrolytic oil with a rate of decomposition of up to 70%, whereas flash pyrolysis maintains a residence period of less than 0.5 s to reduce secondary cracking. On the other hand, the yield of biochar decreases. Additionally, the polymerization reaction in the liquid product is catalyzed by biochar in bio-oil, ensuing in elevated bio-oil viscosity, heightened acidic nature, and diminished thermal strength. These technical constraints are commonly associated with the flash pyrolysis technique [23].

4 Physicochemical Properties of Biochar

The physical characteristics of biochar, particularly its porosity and surface area, have a significant impact on CO_2 capture. The kind of biomass material, the pyrolysis conditions, and the biochar alteration process (physical, chemical, or physicochemical) utilized are the main elements that might alter the surface area and porosity of carbon [24]. Micropores are ideal for CO_2 capture, and excessive pore broadening at high decomposition or activation temperatures would negatively impact the biochar's ability to adsorb gases [12, 25]. Microporosity and specific surface area are two associated characteristics. Creating many tiny micropores would result in a large specific surface area, which would increase CO_2 adsorption. Hence, to get a higher CO_2 adsorptive capability, the porosity of carbon needs to be carefully managed. Biochar's surface chemistry, characterized by different compounds and their functional groups, composition of mineral present, hydrophobic properties, and non-polar nature, and, has a major impact on CO_2 capture.

5 Characterization Techniques for the Biochar

Among the most popular techniques for figuring out the characteristics of biochar is proximate and elemental analysis. The objective of the proximate investigation of biochar is to ascertain its volatile matter, fixed carbon, moisture, and ash contents. High ash content is often not desired since the inorganic mineral substances have the potential to obstruct pores and reduce the approachability of adsorption spots [26]. However, the alkali matter of biochar-ash, such as K_2O and CaO, has been shown by Zubbri et al. [25] to have a positive effect on enhancing surface basic nature and contributing to the capture of CO_2 (with an acidity nature) through carbonate production. Other major characterization techniques that characterize the properties of pristine and engineered biochar for environmental application are depicted in Fig. 4.

6 Application of Biochar in Other Areas

Because lignocellulosic leftovers and organic wastes are widely available and abundant, it is possible to synthesize porous carbon adsorbents from them on an extensive basis and at an affordable price. As seen in Fig. 5, biochar, an instantaneous carbon-based solid substance derived using different techniques, has possibilities in environmental treatment such as adsorption of various pollutants originated from different sources and its direct use [15, 27–31].

7 Modification Techniques of Biochar Surface

Among various adsorbents, the prevalent solid form of adsorbent used in the pressure swing adsorption (PSA) unit is activated carbon [32]. Heat processing of carbon-based substances such as biomass waste, industrial waste, and household waste is typically included in the making of activated carbon (AC). Activation stages include both chemical and physical modification. Numerous chemical activation compounds exist, including potassium hydroxide (KOH), sodium hydroxide (NaOH), phosphoric acid (H_3PO_4), potassium carbonate (K_2CO_3), zinc chloride (ZnCl), and others, however, they are often expensive and harmful to the environment. Physical activation, on the other hand, uses activating substances such as carbon dioxide, steam, or air and is not harmful to the environment.

Physicochemical characteristics of biochar, such as its porous nature, basic nature of the surface, functional groups on its surface, the amount of AAEM, hydrophobicity, orienta-

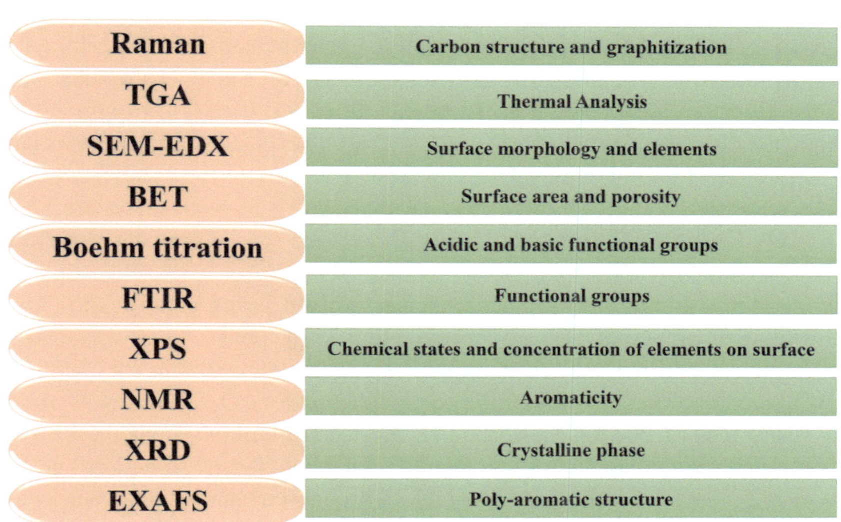

Fig. 4 Characterization methods to analyze the physicochemical properties of biochar

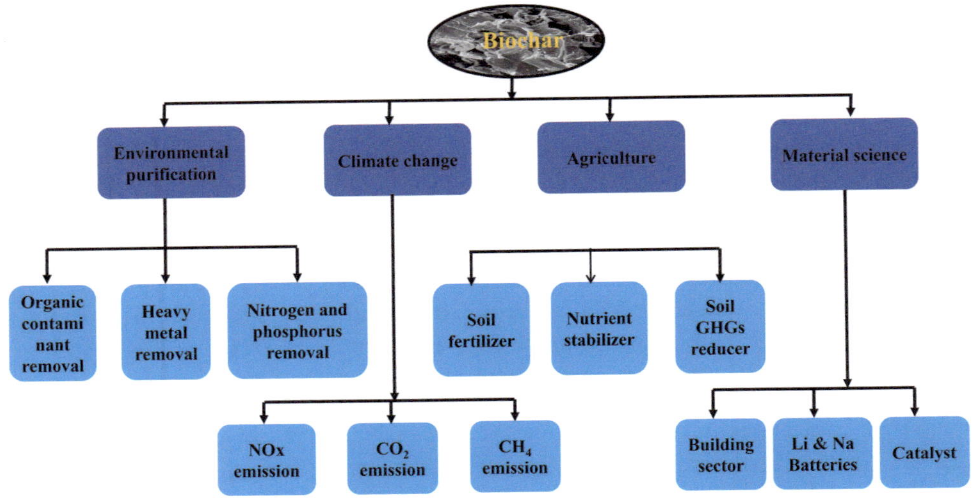

Fig. 5 Diverse applications of biochar across various domains [31]

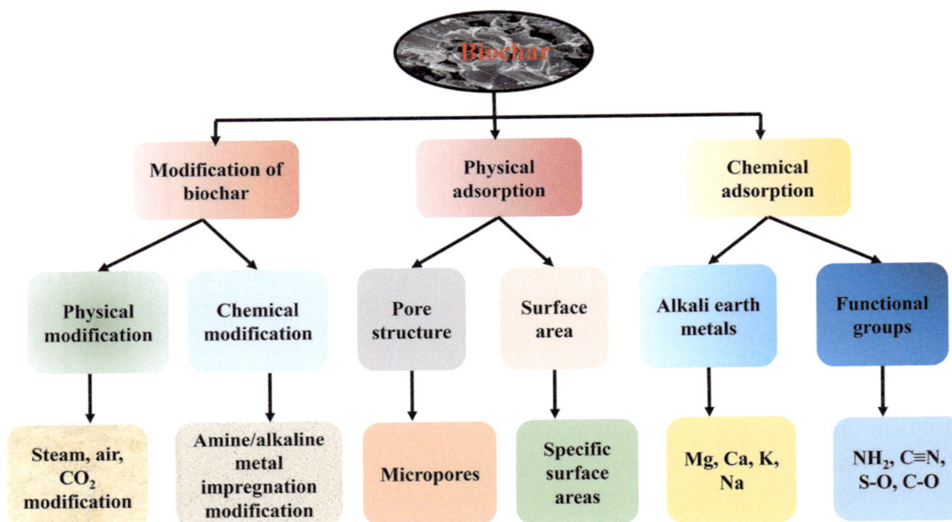

Fig. 6 Physical and chemical alterations, as well as the adsorption mechanism of CO_2 on absorbents derived from biomass [1]

tion, and aromatic properties all influence its ability to adsorb CO_2 [33]. The process of preparation, the origin of carbon, activation, and modifications such as grafting, physicochemical modification, or impregnation with elements as well as the feedstock's pre- and post-treatments all affect these characteristics (Fig. 6).

8 CO_2 Capture Processes Using Biochar

The most notable interactions and mechanism involved during adsorption or capture of CO_2 are shown on Fig. 7. The physical structure and the surface chemistry of biochar dictate how it interacts. The micropore-filling action dominates CO_2 physisorption [12, 34]. CO_2 adsorption is most successful in micropores with a diameter of less than 0.5 nm, while those between 0.3 and 0.8 nm are also suitable. Through the van der Waals interaction, aromatic groups in biochar can also aid in the physisorption of CO_2. Functional groups with an O atom may cause hydrogen bonds to form between the adsorbent and the adsorbate. Since the functional groups containing O are acidic, they would typically prevent acidic CO_2 groups from adhering to them. Nonetheless, some research suggested that the biochar matrix's hydroxyl (–OH) and carboxyl (–COOH) concentrations could improve the hydrogen bonding that occurs between CO_2 molecules [35, 36]. A further mechanism contributing to the adsorption of CO_2 is the Lewis acid–base reaction between basic N-containing functional groups and acidic CO_2 molecules [37].

It was discovered that a variety of interactions discussed in Fig. 7 cause varying adsorption energies with CO_2 in distinct N-containing functional groups [38]. It is confirmed both theoretically and empirically that biochar containing functional groups of pyridone, pyridine, and pyrrole exhibits significant interactions with carbon dioxide [39]. Consequently, to produce engineered biochar with an increased

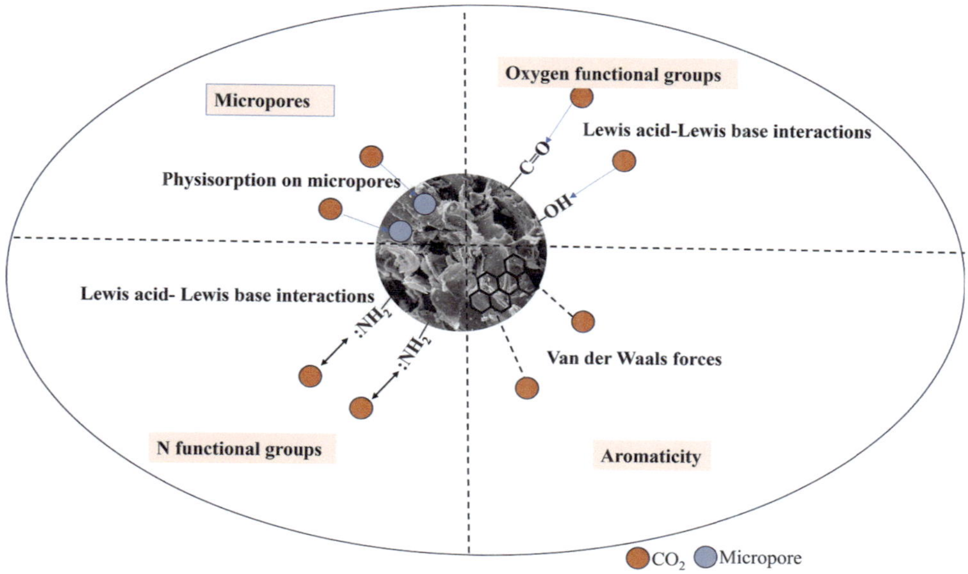

Fig. 7 Potential mechanisms implicated in CO_2 adsorption onto biochar [40]

adsorption capacity at a reduced cost, much study and examination into the surface chemistry and structural characteristics of biochar are required.

9 Development of Engineered Biochar

Low surface area and porosity characterize the textural qualities of pure or unaltered biochar [41, 42]. Furthermore, virgin biochar has minimal basicity, few functional groups, and unsuitable chemical surface properties [43, 44]. Therefore, to create "engineered biochar" with improved CO_2 adsorption capacity and selectivity, one must tweak the surface chemistry of biochar and adjust its inherent features.

Engineered biochar, a modified/functionalized/ composite biochar form of traditional biochar that is more capable of adsorbing CO_2 and more selective than pure biochar, emerges as a multifaceted solution that not only addresses the need for sustainable waste management but also plays a pivotal role in carbon sequestration. By harnessing the power of biochar, researchers and practitioners aim to develop efficient and scalable methods to capture CO_2 from various sources, including industrial processes, power generation, and atmospheric removal. The physicochemical properties of engineered biochar should be advantageous for the adsorption of CO_2 via physical and chemical adsorption [45].

Although research has demonstrated that a moderately sized surface area with increased surface basicity can lead to substantial CO_2 adsorption, it's important to note a higher area of surface does not necessarily assurance a correspondingly higher CO_2 adsorption capability, as indicated by several studies [12]. Furthermore, the designed biochar employed for this purpose needs to be stable during several rounds of CO_2 capture and desorption and exhibit a high degree of choosiness for CO_2 among different gases such as nitrogen, moisture and air [25]. Therefore, to guarantee the higher choosiness and CO_2 capture of biochar, it is required to modify its porosity and surface chemistry. To create designed biochar for CO_2 capture physicochemical modifications should be applied.

9.1 Physical Modification

Physical alteration of biochar is easier and more cost-effective than using costly chemicals. Biochar is often physically modified by applying an oxidizing chemical to it together with a suitable amount of steam, ozone, carbon dioxide, or air [46]. The physical processing of biochar often entails heating it in an active environment or oxidative agent (most commonly, vapor and CO_2) [40, 47].

There are two ways to achieve physical activation: either activate biomass directly with an oxidizing agent or activate biochar by employing an oxidative agent following pyrolysis in the limited oxygen or air. The oxidizing agent enters its matrix of carbon at a higher temperature (> 700 °C) and eliminates few carbon elements, generating other pores and widening the old ones. The primary goal of physically activate is to enhance the inner specific surface area and pores of biochar [46].

When it comes to industrial-scale applications, physical treatment employing steam or CO_2 is more practical than chemical treatment since it uses no chemicals, making it eco-friendly, less expensive, prevents additional contamination, saves time, and yields biochar with lower contaminants [48].

9.1.1 Activation with Carbon Dioxide

Biochar activation with carbon dioxide is most often utilized physical activation procedures for creating tailored biochar. Biochar micropores are enhanced by CO_2 activation, which encourages CO_2 adsorption [1]. CO_2 experiences dissociative chemisorption on the surface of the biochar during the CO_2 activation process, resulting in the formation of surface oxides and CO. The pore is then further developed by the oxides being removed from the surface.

9.1.2 Activation Using Steam

Another common alteration technique is activation with steam, which can help with degradation, the formation of a crystalline phase, and the removal of impurities such as insufficient combustion products [49]. The oxygen of water vapor is transferred to the free active sites on the biochar surface throughout the procedure of alteration, and the hydrogen generated joins the carbon on the biochar's surface to form surface hydrogen complexes [50]. In comparison to CO_2-activated biochar, steam-activated biochar has a larger overall pore volume. This disparity may be explained by the smaller size of water molecules compared to CO_2, which aids in the penetration of steam into the pores and the generation of a higher number of oxygen-containing functional groups on the biochar surface. Consequently, steam-activated biochar tends to be more effective for the purposes CO_2 remediation [51].

$$C + H_2O \rightarrow H_2 + CO \quad (1)$$

$$2C + H_2 \rightarrow 2CH \quad (2)$$

$$H_2O + CO \rightarrow H_2O + CO_2 \quad (3)$$

9.2 Chemical Modification

To enhance the virgin biochar's surface chemical characteristics, such as its surface basicity, functional groups on the surface, polarization, hydrophobicity, and aromaticity, chemical modification is necessary [45]. Chemical activation can be achieved in one of two ways: either by directly soaking biomass with the chemical agent and then thermally treating it, or by soaking the pyrolytic char using chemical agent and then thermally treating it [46, 52]. One way to chemically treat biochar is to oxidize it with the solutions having basic in nature such as potassium and sodium hydroxide etc. [53, 54]. Biochar can be amine-treated by adding ammonia or other amine group [55, 56]. Alternatively, biochar can be impregnated with a metal oxide solution to incorporate basic metals into the structure of the carbon. In chemical treatment, three primary modifications stand out: alkali modification, amine modification, and impregnation with metals or metal oxides. These approaches are widely acknowledged as the most commonly employed methods to enhance the surface chemistry of char surfaces [7].

9.2.1 Alkali-Modified Biochar

There has been a lot of interest in alkali-treating char with regular basic solutions, including potassium and sodium hydroxide, to improve the physicochemical characteristics of virgin biochar. Here, the CO_2 absorption capacity is greatly improved above that of physical treatment because both the chemical (surface chemistry) and physical (porosity) qualities are altered. Base-treated carbon has the benefit of having greater surface basicity, which increases its ability to interact with the acidic CO_2 gas. By splitting and breaking down the graphitic layers of biochar, alkali modification with KOH and subsequent heat treatment causes subsequent reduction and oxidation processes that result in the formation of micropores and mesopores [46]. KOH activation results in biochar with increased phenolic, carboxylic, aromatic, and alkene compounds in the surface functional groups. However, the temperature of activation affects their existence since these functional groups break down into CO and CO_2 at varying degrees. KOH treatment was shown to boost the surface basicity of biochar for CO_2 adsorption purposes because it increased the number of functional groups that contain oxygen [7].

9.2.2 Amino-Modified Biochar

A shift in the adsorption process from physical adsorption to chemical adsorption or mixed physisorption occurs when amines are added to biochar, and at very high temperatures, this can improve the ability to select CO_2. Urea compound has been commonly utilized to add N_2 onto char surfaces due to its safety and cost-effectiveness as an amine source. Surface modification with amino reagents allows for the attachment of relative groups to the material to increase the CO_2 capture capabilities.

The surface basicity of biochar can be enhanced by various N-containing functional groups. By employing N-containing precursors including melamine, polyacrylonitrile, NH_3, amines, and HNO_3 onto the surface of biochar. One popular technique for adding amino groups to the surface of biochar is surface amination, which may be carried out in three different ways: Nitration may be done in three different ways: (1) with NH_3, (2) using reagents that include amino acids, including amines, and (3) by combining the nitration and reduction processes [46].

9.2.3 Oxide Doping of Metal

To improve the pure biochar's ability to capture CO_2 and selectivity, oxide of metals or only metal can be added. Solutions of metal salts including aluminum chloride, ferrous chloride and magnesium chloride can be used for this purpose and then impregnated into biochar. The metallic salt can then be thermally treated to turn it into metal oxide.

10 Functionalization of Biochar Surface

Functionalizing the surface of biochar is a favorable approach for CO_2 adsorption. By introducing identifiable functional groups on top of the biochar, their affinity for CO_2 can be increased, directing to enhanced adsorption efficiency. Various methods can be employed to functionalize biochar, including: (i) Amine functionalization (introducing amine groups (-NH_2) onto the biochar surface can significantly enhance CO_2 adsorption due to the strong interaction between CO_2 and amine functional groups, (ii) Carboxyl functionalization (incorporating carboxyl groups (-COOH) onto biochar can also improve CO_2 capture by forming carboxylate-CO_2 complexes), (iii) Hydroxyl functionalization (Hydroxyl groups (-OH) can be added to the biochar surface, enhancing its polarity and thereby increasing its CO_2 adsorption capacity), (iv) Pyrolysis with nitrogen-containing precursors (During the pyrolysis process, nitrogen-containing precursors such as ammonia (NH_3), amines, or nitrogen-rich organic compounds can be used to introduce nitrogen-functional groups onto the biochar surface, which enhances CO_2 adsorption), (v) Surface modification with other functional groups (Various other functional groups such as imines, imides, and amidoxime can also be employed to modify the biochar surface for improved CO_2 capture). Moreover, the functionalization of biochar surface for CO_2 capture involves tailoring its chemical composition to increase the presence of specific functional groups that have a high affinity for CO_2, thereby enhancing its performance as a CO_2 adsorbent in carbon capture and storage applications.

10.1 Oxygen-Containing Functional Groups

The oxygen-containing functional groups' pyrones, chromenes, and ketones add to the surface's alkalinity. The surface of carbon and CO_2 is believed to help form hydrogen bonds more easily. This is especially true when oxygen atoms are introduced. The introduction of oxygen atoms leads to the formation of functional groups having acidic oxygen-contains. These groups include carboxyl and hydroxyl groups. Adsorbent polarity is increased by carboxyl and hydroxyl groups, which helps CO_2 adsorption with the quadrupole moment. According to Liu et al. [57], the presence of oxygen-containing functional groups enhanced the amount of CO_2 adsorbed in the microporous structure because more CO_2 flows to areas with greater electron densities surrounding these groups.

10.2 Nitrogen-Containing Functional Groups

Functionalization of biochar with nitrogen-containing functional groups is a promising strategy for various applications, including CO_2 capture. Nitrogen-containing functional groups can be added to the biochar surface. This can be done using various methods. One method is animation. Another method involves pyrolysis with nitrogen-rich precursors. Chemical modification is also a technique used for this purpose. The introduction of nitrogen-containing functional groups onto biochar enhances its surface properties, including surface area, pore structure, and surface chemistry. These modifications can improve biochar's performance in various applications, such as CO_2 capture, pollutant removal from water and soil, and nutrient retention in agricultural soils. Additionally, nitrogen-functionalized biochar can exhibit increased adsorption capacity and selectivity for target molecules due to specific interactions between nitrogen sites and adsorbates [32].

10.3 Sulfur-Based Functional Groups

Functionalization of biochar with sulfur-containing functional groups represents a promising strategy with diverse applications, including environmental remediation, energy storage, and catalysis. Functionalization with sulfur-containing compounds has not yet been studied in connection to CO_2 adsorption, in contrast to those containing nitrogen [58]. Biochar can be modified using the sulfonation method by treating biochar with sulfuric acid (H_2SO_4) or sulfur dioxide (SO_2) under controlled conditions. This process leads to the incorporation of sulfonic acid (-SO_3H) or other sulfur-containing groups onto the biochar surface. Biochar can be chemically modified using sulfur-containing reagents such as thiol compounds (R-SH) or sulfur-containing organic compounds. Moreover, during the pyrolysis process, sulfur-rich compounds such as thiophene, thioacetamide, or sulfur-containing polymers can be used as precursors. These compounds decompose during pyrolysis, leaving sulfur-functionalized groups on the resulting biochar [59].

11 Influence of Process Parameters on CO_2 Adsorption

Because different biomass feedstocks have different structures and compositions, as well as different pyrolysis reaction parameters, biochar's structure and physiochemical characteristics change. To enhance and control the properties of biochar, it is important to assess operational parameters. These parameters include pyrolysis and gasification conditions. Evaluating their impact helps determine how they influence the material's adsorption capacity.

11.1 Effect of Feedstocks

The amount of CO_2 adsorbed on the biochar that is made from various virgin biomass materials varies. The type of feedstock used influences the chemical and physical properties of biochar. A high number of heteroatoms, such as nitrogen (N) and sulfur (S), are present on the surface. These heteroatoms contribute to a significant increase in CO_2 adsorption.

11.2 Effect of Temperature

The temperature at which biochar is pyrolyzed also affects its structural and compositional characteristics, including surface area, pore structures, surface functional groups, and component compositions. Biochars have an adsorption capacity that rises progressively with increasing temperature up to a specific point. This is due to the low temperature-produced biochar's having poor porosity, low specific surface area, and susceptibility to tar generated during biomass pyrolysis [60]. The increase in carbon content with rising temperatures offset the reduction in nitrogen content, which helped to boost CO_2 adsorption capability. When the pyrolysis temperature increased to 800 °C, the capture capacity decreased. The values were much lower compared to those obtained at 700 °C. This may be explained by the decreasing nitrogen concentration and the surface porosity being destroyed at a higher temperature (800 °C).

11.3 Effect of Heating Rate

The heating rate has a major impact on biochar output. The yields of biochar decreased as the heating rate increased [61]. High heating rates lead to the elimination of structural complexity because of phase transitions, swelling, and localized melting of cell structures. Low heating rates help to preserve structural complexity. Lower ramping is typically preferred for biochar production. Although they take longer, they offer better yield and durability.

11.4 Effect of Residence Time

According to research done by Yaashikaa et al. [62], biochar output increases with longer residence durations at lower pyrolytic temperatures. This is largely caused by the condensation and polymerization processes that take place between the solids and vapors generated.

12 Future Directions

No technological solution is without its challenges. This chapter included a thorough discussion of the principles underlying the several modification strategies used to produce engineered biochar with increased CO_2 adsorption capability. According to earlier research, the tailored biochar's chemical and physical characteristics affect CO_2 adsorption. However, further research is needed to achieve desired characteristics. The engineered biochar can capture high amounts of CO_2 and remain stable during the process of regeneration. On the other hand, its use in CO_2 capture is currently limited and has several drawbacks.

Further investigation is needed to determine if the components of different biochar forms interact. It is also important to understand whether biochar interacts with the feedstocks during the adsorption process. This occurs when biochar is produced through co-pyrolysis of multiple feedstocks. Moreover, future research may examine if producing biochar with a high adsorption capacity. Apart from the conventional physical and chemical changes, some newly developed modification techniques can be used to modify the surface functional groups and surface area of biochar. These techniques include ultrasonic, microwave, plasma, and electrochemical modification. As a result, further techniques of modification may be created. Because of its stability, biochar lasts a long period in the environment. Thus, a crucial area of research is the production of biochar composites that are harmless to the environment.

Furthermore, most studies have reported the use of modified biochar for CO_2 capture in the laboratory setting in which pure gases, typically a given mixture of N_2 and CO_2, were used as the feed gas. The effect of moisture and other impurities such as SOx and NOx must be investigated to test the durability and long-term use of modified biochar for CO_2 adsorption ability.

13 Conclusion

The use of engineered biochar material for CO_2 adsorption presents a promising avenue in the quest for sustainable carbon capture technologies. Because of its natural characteristics, biochar can be a sustainable and affordable CO_2

capture material. This book chapter examines the several methods used to produce biochar as well as the mechanism by which it adsorbs CO_2. Through various methods such as functionalization with nitrogen-containing groups, biochar can be tailored to exhibit enhanced CO_2 adsorption properties. The versatility of biochar production methods and its abundance as a byproduct of biomass conversion processes further contribute to its attractiveness as a carbon capture material. Moreover, the utilization of biochar for CO_2 capture not only mitigates greenhouse gas emissions but also offers potential synergies with waste management and agricultural practices. However, further research is needed to optimize biochar properties, scale up production, and assess its long-term environmental impacts. With continued innovation and investment, engineered biochar holds significant promise as a cost-effective and environmentally sustainable solution for CO_2 capture and mitigation in the fight against climate change.

Nomenclature

KOH	Potassium hydroxide
CCUS	Carbon, capture, utilization, and storage
IPCC	Intergovernmental Panel on Climate Change
PPM	Parts per million
GHG	Greenhouse gas
Mtpa	Million metric tons per annum
CAGR	Compound annual growth rate
IEA	International Energy Agency
PSA	Pressure swing adsorption
AC	Activated carbon

References

1. Guo S, Li Y, Wang Y, Wang L, Sun Y, Liu L (2022) Recent advances in biochar-based adsorbents for CO_2 capture. Carbon Capture Sci Technol 4:100059. https://doi.org/10.1016/j.ccst.2022.100059
2. Qiao Y, Zhang S, Quan C, Gao N, Johnston C, Wu C (2020) One-pot synthesis of digestate-derived biochar for carbon dioxide capture. Fuel 279:118525. https://doi.org/10.1016/j.fuel.2020.118525
3. Raganati F,Miccio F, Ammendola P (2021) Adsorption of carbon dioxide for post-combustion capture : a review. https://doi.org/10.1021/acs.energyfuels.1c01618
4. Zhu X, Li S, Shi Y, Cai N (2019) Recent advances in elevated-temperature pressure swing adsorption for carbon capture and hydrogen production 75
5. Arias F, Sen TK (2009) Removal of zinc metal ion (Zn^{2+}) from its aqueous solution by kaolin clay mineral: A kinetic and equilibrium study. Colloids Surfaces A Physicochem Eng Asp 348:100–108. https://doi.org/10.1016/j.colsurfa.2009.06.036
6. Karimi M, Shirzad M (2022) Biomass/Biochar carbon materials for CO_2 capture and sequestration by cyclic adsorption processes: a review and prospects for future directions. J CO_2 Util 57. https://doi.org/10.1016/j.jcou.2022.101890
7. Natasha A, Rahman A, Lahijani P, Mohammadi M (2021) Recent advances in developing engineered biochar for CO_2 capture : an insight into the biochar modification approaches. J Environ Chem Eng 9:106869. https://doi.org/10.1016/j.jece.2021.106869
8. Mukherjee A, Chatterjee S (2019) Capture utilization and storage (CCUS)—policy framework and deployment mechanism in India. NITI Aayog Report 2019
9. Lee Y, Lee SW, Tsang YF, Kim YT, Lee J (2020) Engineered rice-straw biochar catalysts for the production of value-added chemicals from furan. Chem Eng J 387:124194. https://doi.org/10.1016/j.cej.2020.124194
10. Premchand P, Demichelis F, Chiaramonti D, Bensaid S, Fino D (2023) Biochar production from slow pyrolysis of biomass under CO_2 atmosphere: a review on the effect of CO_2 medium on biochar production, characterization, and environmental applications. J Environ Chem Eng 11:110009. https://doi.org/10.1016/j.jece.2023.110009
11. Xue Y, Guo Y, Zhang X, Kamali M, Aminabhavi TM, Appels L, Dewil R (2022) Efficient adsorptive removal of ciprofloxacin and carbamazepine using modified pinewood biochar–a kinetic, mechanistic study. Chem Eng J 450:137896. https://doi.org/10.1016/j.cej.2022.137896
12. Lahijani P, Mohammadi M, Rahman A (2018) Metal incorporated biochar as a potential adsorbent for high capacity CO_2 capture at ambient condition 26:281–293
13. Kan T, Strezov V, Evans TJ (2016) Lignocellulosic biomass pyrolysis: a review of product properties and effects of pyrolysis parameters. Renew Sustain Energy Rev 57:1126–1140. https://doi.org/10.1016/j.rser.2015.12.185
14. Ouyang J, Hong D, Jiang L, Li Z, Liu H, Luo G, Yao H (2020) Effect of CO_2 and H_2O on char properties. Part 1: pyrolysis char structure and reactivity. Energy Fuels 34:4243–4250. https://doi.org/10.1021/acs.energyfuels.0c00032
15. Wang JX, Cao JP, Zhao XY, Liu TL, Wei F, Fan X, Zhao YP, Wei XY (2017) Study on pine sawdust pyrolysis behavior by fast pyrolysis under inert and reductive atmospheres. J Anal Appl Pyrolysis 125:279–288. https://doi.org/10.1016/j.jaap.2017.03.015
16. Lee XJ, Ong HC, Gan YY, Chen WH, Mahlia TMI (2020) State of the art review on conventional and advanced pyrolysis of macroalgae and microalgae for biochar, bio-oil, and bio-syngas production. Energy Convers Manag 210. https://doi.org/10.1016/j.enconman.2020.112707
17. Zaini IN, Sophonrat N, Sjöblom K, Yang W (2021) Creating values from biomass pyrolysis in Sweden: Co-production of H2, biocarbon and bio-oil. Processes 9:1–21. https://doi.org/10.3390/pr9030415
18. Buffi M, Rizzo AM, Pari L, Chiaramonti D (2018) Intermediate pyrolysis campaign to assess products yield and quality varying biomass particles size, temperatures, and composition. Int J Oil, Gas Coal Technol 17:355–373. https://doi.org/10.1504/IJOGCT.2018.090146
19. Hu Y, Gong M, Feng S, (*Charles*) Xu C, Bassi A (2019) A review of recent developments of pre-treatment technologies and hydrothermal liquefaction of microalgae for bio-crude oil production. Renew Sustain Energy Rev 101:476–492. https://doi.org/10.1016/j.rser.2018.11.037
20. Mašek O (2016) Biochar in thermal and thermochemical biorefineries-production of biochar as a coproduct. In: Handbook of biofuels production processes and technologies. Second Ed 655–671. https://doi.org/10.1016/B978-0-08-100455-5.00021-7

21. Mulabagal V, Baah DA, Egiebor NO, Chen W (2020) Handbook of climate change mitigation and adaptation. https://doi.org/10.1007/978-1-4614-6431-0
22. Li J, Lv G, Bai W, Liu Q, Zhang Y, Song J (2016) Modification and use of biochar from wheat straw (*Triticum aestivum* L.) for nitrate and phosphate removal from water 57:19443994. https://doi.org/10.1080/19443994.2014.994104
23. Cornelissen T, Yperman J, Reggers G, Schreurs S, Carleer R (2008) Flash co-pyrolysis of biomass with polylactic acid. Part 1: Influence on bio-oil yield and heating value. Fuel 87:1031–1041. https://doi.org/10.1016/j.fuel.2007.07.019
24. Leng L, Xiong Q, Yang L, Li H, Zhou Y, Zhang W, Jiang S, Li H, Huang H (2021) An overview on engineering the surface area and porosity of biochar. Sci Total Environ 763. https://doi.org/10.1016/j.scitotenv.2020.144204
25. Zubbri NA, Mohamed AR (2020) Enhancement of CO_2 adsorption on biochar sorbent modified by metal incorporation 11809–11829
26. Li S, Harris S, Anandhi A, Chen G (2019) Predicting biochar properties and functions based on feedstock and pyrolysis temperature: a review and data syntheses. J Clean Prod 215:890–902. https://doi.org/10.1016/j.jclepro.2019.01.106
27. Liu WJ, Jiang H, Yu HQ (2019) Emerging applications of biochar-based materials for energy storage and conversion. Energy Environ Sci 12:1751–1779. https://doi.org/10.1039/c9ee00206e
28. Mensah AK, Frimpong KA (2018) Biochar and/or compost applications improve soil properties, growth, and yield of maize grown in acidic rainforest and coastal savannah soils in Ghana. Int J Agron. https://doi.org/10.1155/2018/6837404
29. Dissanayake PD, You S, Igalavithana AD, Xia Y, Bhatnagar A, Gupta S, Kua HW, Kim S, Kwon JH, Tsang DCW, Ok YS (2020) Biochar-based adsorbents for carbon dioxide capture: a critical review. Renew Sustain Energy Rev 119:1–14. https://doi.org/10.1016/j.rser.2019.109582
30. Rahman S, Hossain A, Ahmed MB (2020) Dioxide capture 1–17
31. Dai Y, Zhang N, Xing C, Cui Q, Sun Q (2019) The adsorption, regeneration, and engineering applications of biochar for removal organic pollutants: a review. Chemosphere 223:12–27. https://doi.org/10.1016/j.chemosphere.2019.01.161
32. Shen Y, Linville JL, Urgun-demirtas M, Schoene RP, Snyder SW (2015) Producing pipeline-quality biomethane via anaerobic digestion of sludge amended with corn stover biochar with in-situ CO_2 removal 158:300–309
33. Chiang Y, Juang R (2017) Surface modifications of carbonaceous materials for carbon dioxide adsorption: a review 71:214–234
34. Klasson KT, Uchimiya M, Lima IM (2015) Characterization of narrow micropores in almond shell biochars by nitrogen, carbon dioxide, and hydrogen adsorption 67:33–40
35. Liu Y, Wilcox J (2012) Effects of surface heterogeneity on the adsorption of CO_2 in microporous carbons. Environ Sci Technol 46:1940–1947. https://doi.org/10.1021/es204071g
36. Xing W, Liu C, Zhou Z, Zhou J, Wang G, Zhuo S, Xue Q (2014) Oxygen-containing functional group-facilitated CO_2 capture by carbide-derived carbons 1–8
37. Plaza MG, Pevida C, Martín CF, Fermoso J, Pis JJ, Rubiera F (2010) Developing almond shell-derived activated carbons as CO_2 adsorbents 71:102–106. https://doi.org/10.1016/j.seppur.2009.11.008
38. Lim G, Lee KB, Ham HC (2016) Effect of N-containing functional groups on CO_2 adsorption of carbonaceous materials: a density functional theory approach. https://doi.org/10.1021/acs.jpcc.5b12090
39. Fan X, Zhang L, Zhang G, Shu Z, Shi J (2013) Chitosan derived nitrogen-doped microporous carbons for high-performance CO_2 capture, 1:0–7
40. Igalavithana AD, Choi SW, Shang J, Hanif A, Dissanayake PD, Tsang DCW, Kwon JH, Lee KB, Ok YS (2020) Carbon dioxide capture in biochar produced from pine sawdust and paper mill sludge: effect of porous structure and surface chemistry. Sci Total Environ 739. https://doi.org/10.1016/j.scitotenv.2020.139845
41. Liu W, Jiang H, Yu H (2015). Development of biochar-based functional materials: toward a sustainable platform carbon material. https://doi.org/10.1021/acs.chemrev.5b00195
42. Krasucka P, Pan B, Sik Y, Mohan D, Sarkar B, Oleszczuk P (2021) Engineered biochar—a sustainable solution for the removal of antibiotics from water 405
43. Ding S, Liu Y (2020) Adsorption of CO_2 from flue gas by novel seaweed-based KOH-activated porous biochars 260
44. Wan Z, Sun Y, Tsang DCW, Khan E, Yip ACK, Hau Y (2020) Customised fabrication of nitrogen-doped biochar for environmental and energy applications 401
45. Dulanja P, You S, Deshani A, Xia Y, Bhatnagar A, Gupta S, Wei H, Kim S, Kwon J, Tsang DCW, Sik Y (2020) Biochar-based adsorbents for carbon dioxide capture: a critical review 119:1–14
46. Sajjadi B, Zubatiuk T, Leszczynska D, Leszczynski J, Chen WY (2018) Chemical activation of biochar for energy and environmental applications: a comprehensive review
47. Zhang X, Wu J, Yang H, Shao J, Wang X, Chen Y, Zhang S, Chen H (2016) Preparation of nitrogen-doped microporous modified biochar by high-temperature CO_2-NH_3 treatment for CO_2 adsorption: effects of temperature. RSC Adv 6:98157–98166. https://doi.org/10.1039/c6ra23748g
48. Rashidi NA, Yusup S (2020) Biochar as potential precursors for activated carbon production: parametric analysis and multi-response optimization 27480–27490
49. Upamali A, Chen SS, Tsang DCW, Zhang M, Vithanage M, Mandal S, Gao B, Bolan NS, Sik Y (2016) Chemosphere Engineered/designer biochar for contaminant removal/immobilization from soil and water: potential and implication of biochar modification 148:276–291
50. Wang J, Wang S (2019) Preparation, modification and environmental application of biochar: a review 227:1002–1022
51. Zhao H (n.d.) Review carbon-based adsorbents for post-combustion capture: a review 26:1–26. https://doi.org/10.1002/ghg
52. Boshir M, Zhou JL, Ngo HH, Guo W, Chen M (2016) Progress in the preparation and application of modified biochar for improved contaminant removal from water and wastewater 214:836–851
53. Jin H, Capareda S, Chang Z, Gao J, Xu Y, Zhang J (2014) Biochar pyrolytically produced from municipal solid wastes for aqueous As (V) removal: adsorption property and its improvement with KOH activation 169:622–629
54. Ellis N, Dehkhoda AM (2016) A novel method to tailor the porous structure of KOH-activated biochar and its application in capacitive deionization and energy storage 87:107–121
55. Chatterjee R, Sajjadi B, Mattern DL, Chen W, Zubatiuk T, Leszczynska D, Leszczynski J, Egiebor NO, Hammer N (2018) Ultrasound cavitation intensified amine functionalization: a feasible strategy for enhancing CO_2 capture capacity of biochar 225:287–298
56. Chatterjee R, Sajjadi B, Chen W, Mattern DL, Hammer N (2020) Effect of pyrolysis temperature on physicochemical properties and acoustic-based amination of biochar for efficient CO_2 Adsorption 8:1–18. https://doi.org/10.3389/fenrg.2020.00085
57. Liu Y, Wilcox J (2012) Effects of surface heterogeneity on the adsorption of CO_2 in microporous carbons
58. Saha D, Kienbaum MJ (2019) Role of oxygen, nitrogen and sulfur functionalities on the surface of nanoporous carbons in CO_2 adsorption: a critical review. Microporous Mesoporous Mater 287:29–55. https://doi.org/10.1016/j.micromeso.2019.05.051
59. Saha D, Orkoulas G, Chen J, Hensley DK (2017) Adsorptive separation of CO_2 in sulfur-doped nanoporous carbons: selectivity and breakthrough simulation. Microporous Mesoporous Mater 241:226–237. https://doi.org/10.1016/j.micromeso.2016.12.015

60. Patra BR, Mukherjee A, Nanda S, Dalai AK (2021) Biochar production, activation and adsorptive applications: a review 2237–2259
61. Şensöz S, Angin D (2008) Pyrolysis of safflower (*Charthamus tinctorius* L.) seed press cake in a fixed-bed reactor: Part 2. Structural characterization of pyrolysis bio-oils. Bioresour Technol 99:5498–5504. https://doi.org/10.1016/j.biortech.2007.11.004
62. Yaashikaa PR, Kumar PS, Varjani SJ, Saravanan A (2019) Advances in production and application of biochar from lignocellulosic feedstocks for remediation of environmental pollutants 292
63. Liu C, Xing W, Zhou J, Zhuo SP (2013) N-containing activated carbons for CO_2 capture. Int J Smart Nano Mater 4:55–61. https://doi.org/10.1080/19475411.2012.668861

Carbon Dioxide Capture and Utilisation in Steel Industry

Raghavendra Ragipani and Anshika

1 Introduction

Construction materials, such as steel and cement, contribute substantially to global carbon dioxide emissions. Prospects of sustainable infrastructure, including energy generation and storage, is intertwined with the availability of low-carbon materials such as green steel [1]. However, decarbonising industries such as iron and steel mills are considered difficult for the following reasons and often termed as hard-to-abate industries. Currently, iron & steel industries are dependent on carbon-rich material inputs, thereby directly contributing 7–8% to global CO_2 emissions and a total of about 10% when emissions associated with electricity consumption are also considered [2]. As such, iron and steel mills are the largest industrial consumers of the coal [2]. Carbon dioxide emissions associated with steelmaking can be classified into combustion and process emissions [3]. Combustion emissions are associated with use of fuels for energy production. On the other hand, process emissions are non-combustion emissions due to chemical reactions necessary for steelmaking. Currently, there are no commercial technologies that allow alternative low-carbon feedstock to fulfil process requirements. Lowering of CO_2 emissions is expected out of improvements in process efficiency, recycling metal, use of hydrogen as a reductant and end-of-pipe carbon capture with utilisation or storage. This chapter briefly reviews carbon dioxide capture and mineralisation opportunities within steel mills.

R. Ragipani (✉) · Anshika
Department of Chemical Engineering, Indian Institute of Technology Kanpur, Kanpur, India
e-mail: ragipani@iitk.ac.in

Anshika
e-mail: anshika21@iitk.ac.in

2 Brief Overview of Steelmaking Processes

Largely, steelmaking is done via two routes—blast furnace and basic oxygen furnace (BF-BOF) and electric arc furnace (EAF). Blast furnaces (BF) are ultra large reactors where iron ore is reduced at high temperature ($\sim$1700 °C) to obtain hot metal. This reduction process currently uses fossil carbon sources such as coal/coke/natural gas. Hot metal from the BF is treated in basic oxygen furnace to obtain crude steel. Although, scrap steel may also be fed, integrated steel mills typically produce virgin steel, which has superior characteristics owing to lower alloying impurities. On the other hand, electric steelmaking via an Electric Arc Furnace (EAF) route processes recycle scrap steel and direct reduced iron (DRI). DRI, also referred to as sponge iron, is produced by direct solid-state reduction of iron ore, typically in rotary kiln, at temperatures below melting point (900–1050°C). Process flow diagrams for BF-BOF and DRI-EAF steelmaking are shown in Fig. 1.

Globally, basic oxygen furnace (BOF) steelmaking is the dominant route with a production contribution of about 72%.[1] Regionally, steelmaking is very heterogeneous. For instance, in India, which is the second largest producer, about 46% of crude steel is produced via BOF, 31% through Induction furnace (IF) and the remaining 23% using Electric Arc Furnace [4]. In the United States, which is the second largest consumer of steel, about two-thirds is produced via scrap recycling in an EAF and remaining via BOF. Given these differences, generation of carbon dioxide and decarbonisation efforts are non-uniform across the globe.

[1] [Accessed on 15 Mar 2024] https://worldsteel.org/steel-topics/sustainability/sustainability-indicators-2023-report/.

S. Narayanan et al. (eds.), *Materials, Devices and Systems for Sustainability*, IITK Directions 8,
https://doi.org/10.1007/978-981-96-7295-0_8

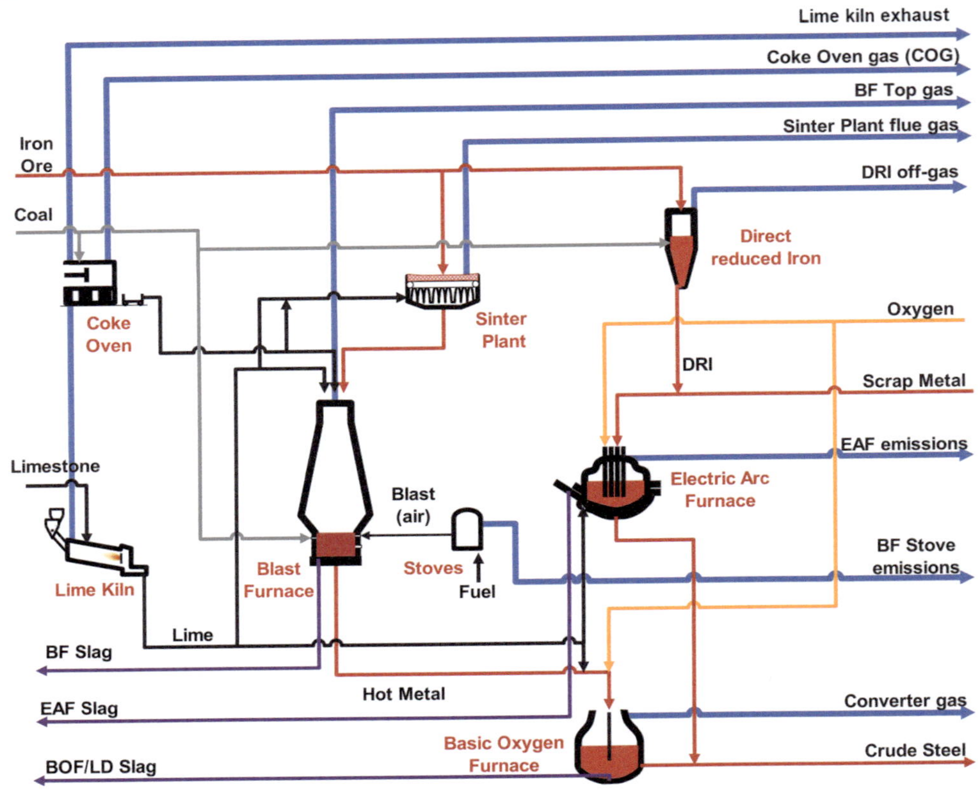

Fig. 1 Process flow sheet showing integrated steel mill (BF-BOF) and coal-based DRI with EAF. CO_2-rich streams are shown in blue lines

3 CO_2 Emissions and Net-Zero Targets by Steel Industry

A comprehensive definition for net-zero, carbon-neutrality and carbon negative can be understood based on atmospheric carbon flux ratio (R_{AF}), which is defined as follows [5]:

$$R_{AF} = \frac{\text{Net anthropogenic } CO_2 \text{ emissions}}{\text{net } CO_2 \text{ uptake by natural sinks}} \tag{1}$$

An atmospheric carbon flux ratio (R_{AF}), of 1 suggests that the net anthropogenic CO_2 emissions are balanced by the uptake by natural sinks such as plants, soil and oceans. Thus, there is no further accumulation of CO_2 in the atmosphere and defines the carbon-neutrality. On the other hand, $R_{AF} = 0$ implies net-zero scenario where the net carbon dioxide emissions, which is the total man-made CO_2 emissions minus artificial carbon sinks, are zero. Similarly, net-zero greenhouse gas (GHG) emissions target includes carbon dioxide as well as other greenhouse gases such as methane. Carbon-negative should be interpreted as negative atmospheric flux ratio ($R_{AF} < 0$). An alternative way to understand the decarbonisation targets, especially those set by the industries, would be based on life cycle assessment (LCA). Total greenhouse gas emissions intensity of any product or service would include all the emissions accumulated from mining, transportation, processing and consumption to final disposal back into environment. The scope of such an analysis is beyond the process boundaries, where the good/service is produced, and is referred to as cradle-to-grave. The total emissions are categorised into—scope 1, scope 2 and scope 3 emissions. Scope 1 refers to the process emissions originating from the production facility. Scope 2 emissions are those associated with purchase of energy or electricity. Scope 3 emissions are all other emissions originating from the entire supply chain. To determine the life cycle inventory, practitioners often depend on carbon emission factors and intensity. Carbon dioxide emission factor for any material or energy input is defined as the mass of CO_2 emitted per unit mass of material or energy usage, typically during an industrial activity. Carbon intensity of a material is the mass of carbon dioxide (equivalent) released into the atmosphere on producing a unit mass of the desired product. Carbon intensity or emission intensity provides an estimate of embodied CO_2 emissions. For a detailed introduction to the LCA and its application to carbon capture and utilisation, the reader is referred to a recently published tutorial review [6]. Global steelmakers have set a short-term target to achieve about 25–35% reduction in emissions by 2030 compared to the mid-2010s. These emission reductions are targeted within scopes 1 and 2. By 2050, most steel industries aim to be carbon-neutral. Although the emission reduction targets of major steelmakers are similar, heterogeneity

Table 1 Emission factors associated with iron and steelmaking and overall carbon intensity of steel

Process	Upstream emission factors (for Iron making)	Direct emissions (Scope 1) for an integrated steelmaking	Carbon intensity of steel (Scope 1, 2 & 3)
Units	tCO_2/t of pig iron or DRI	tCO_2/t of crude steel	tCO_2/t of steel cast
BF-BOF	1.85	2.19	2.33
DRI-EAF (Gas)	0.78	–	1.37
DRI-EAF (coal)	1.2	2.95	3.2
Scrap-EAF	0	0.18	0.68
Global average	–	–	1.91

Data sources: World Steel sustainability indicators report ([Accessed on 15 Mar 2024] https://worldsteel.org/steel-topics/sustainability/sustainability-indicators-2023-report/); Refs [1, 7, 8]

Table 2 Typical composition of CO_2-rich streams from iron and steel mills

Off gas	CO_2	CO	CH_4	CxHy	H_2	N_2	Carbon content	Ref
Lime kiln vent	30	–	–	–	–	70	0.11†	[11]
Sinter	5–10	0–1	–	–	–	70	0.04†	[11]
Coke oven gas	1–5.4	3.4–5.8	15.7–27	1.4–2.4	36.1–61.7	1.5–6	0.47	[12]
BF gas	16–26	19–27	–	–		44–58	0.17	[12]
BOF gas	17.2	60.9	0.1	–	4.3	15.5	0.35	[12]
EAF gas	20	72.9	0	0	5	1.8		[13]

† Estimated

in the steel production routes across the world meant that the decarbonisation roadmaps are non-uniform. Decarbonisation roadmap can be summarised into six pathways, each contributing to a fraction of emissions reduction. These include (i) increased scrap recycling, (ii) shift towards electric arc furnace powered by renewable electricity, (iii) H_2 injection into blast furnace, iv) H_2-based shaft reduction, (v) blast furnace (or direct reduction of iron) with carbon capture and utilisation and (vi) carbon capture storage including blue carbon and afforestation. The rationale for the choice of decarbonization pathway can be understood based on emission factors associated with steel production (Table 1).

The global average carbon dioxide emission intensity for crude steel production is estimated to be 1.77 tCO_2/tcs (Scope 1 & 2) and 1.91 tCO_2/tcs (Scope 1, 2 and 3). As shown in Table 1, carbon intensity of crude steel varies substantially based on the source of reducing gas and steelmaking process. For instance, virgin steel produced from BF-BOF has higher carbon intensity compared to DRI-EAF. Crude steel obtained by recycling scrap steel has much lower carbon intensity. It is apparent that switching from BF-BOF to DRI-EAF and increasing the scrap recycling are prominent objectives. However, given the scale and significance of BF-BOF in producing about 72% of global steel, it is necessary to finds ways to decarbonise it. Crude steel produced via BF-BOF route has an average Scope 1 CO_2 emissions intensity of 2.19 tCO_2/tcs, and overall carbon intensity of 2.33 tCO_2/tcs (Scope 1, 2 and 3). A significant fraction (up to 80%) of these emissions originates in pre-steelmaking units such as sinter plant, coke oven and blast furnace. As shown in the process flow diagram (Fig. 1), coal and coke fed to pre-steelmaking units such as coking, sintering and blast furnace contribute to these emissions. CO_2-rich gaseous streams from these units such as coke oven gas, BF top gas and sinter plant flue gas may be routed for energy recovery via combustion before venting out to atmosphere. Typical composition of these gaseous streams is provided in Table 2. Currently, partial replacement of coke with hydrogen gas injection, and end-of-pipe carbon dioxide capture units are being pursued. Estimates suggest that BF with carbon dioxide capture unit has a potential to reduce overall emissions by 25%. Further emissions reduction, up to 80%, is possible by ‘break-through’ smelting reduction technologies such as HIsarna with carbon capture [9, 10].

4 Carbon Capture in Steel Industry

From the above discussion on decarbonisation roadmaps, carbon capture seems inevitable to achieve net-zero target. Carbon capture units may be deployed either as a start-of-pipe or an end-of-pipe solution, as shown in Fig. 2. A start-of-pipe CO_2 capture unit is a gas scrubbing unit to remove CO_2 from syngas (a mixture of CO and H_2) or hydrogen produced from a carbon-rich material. Syngas or hydrogen may be produced by either gas reforming or coal/biomass gasification. Removal of carbon dioxide by scrubbing improves the hydrogen to carbon ratio (H/C) and reduces the emission factor. As shown in Fig. 2a, reducing gas for a blast furnace can be obtained from a coal gasifier with a CO_2 capture unit. On the other hand,

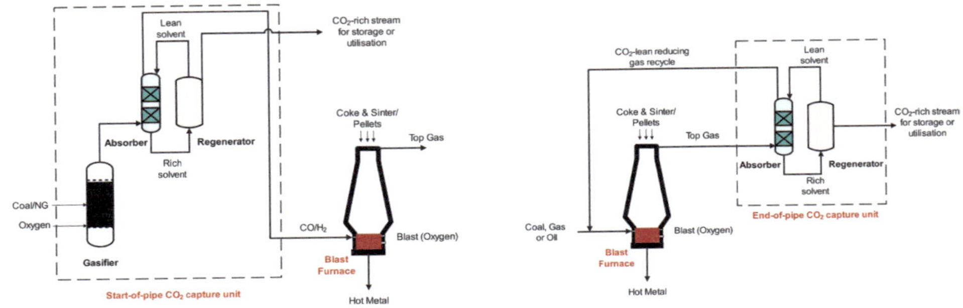

(a) Start-of-pipe (b) End-of-pipe

Fig. 2 Absorption-based CO_2 capture unit for reducing BF CO_2 emissions

an end-of-pipe solution, as the name suggests, is deployed as a scrubber to treat tail gas streams before they are vented to atmosphere. A conventional blast furnace with air blast generates BF gas that is rich in CO, CO_2 and N_2 (see Table 2 for composition). A carbon capture unit may be installed at the downstream of blast furnace to capture CO_2 from the BF gas (16–26 vol% carbon dioxide). However, owing to high concentrations of CO (19–27 vol%), the emission factor for the scrubbed gas is still high. Moreover, because of high N_2 concentration, it is not suitable for recycling to BF. To overcome these challenges, a conventional BF is modified to oxygen-BF as shown in Fig. 2b. Use of oxygen blast and a CO_2 gas scrubbing unit, produces BF gas that has little N_2 and CO_2. Thus, CO-rich gas is fully recycled to blast furnace; such a system is referred to as top gas recycling blast furnace (TGR-BF). Carbon dioxide separated from these units can be converted to fuels and chemicals or sequestered.

Top gas recycling blast furnace (TGR-BF) was proposed as part of European Union's ULCOS program aimed at identifying low-carbon technologies [11]. It is estimated that the CO_2 emissions avoided by TGR-BF is as high as 57% [14]. Moreover, existing blast furnaces can be retrofitted as TGR-BF. Readers are referred to the work by [11], where an extensive discussion on TGR-BF is presented along with other low-carbon technologies for steel mills.

Carbon capture technologies under consideration: Conventionally, amine solvent-based gas scrubbing units are considered for carbon dioxide capture. Typical amine solvents in use are MEA, MDEA, AMP and piperazine [15]. In a scrubbing unit (Fig. 3), there are two steps—absorption and stripping. In the absorption step, gaseous carbon dioxide is contacted with the liquid solvent over packings. CO_2 from the gas phase is absorbed into the solvent and converted to ions such as bicarbonate, carbonate, carbamate ions and impurities in the gas phase leave the absorber. Absorption is typically carried out close to atmospheric temperature. In the stripping step, the temperature of the CO_2-rich solvent containing bicarbonate, carbonate and carbamate ions is increased close to 100 °C such that carbon dioxide is separated from the solvent. Vacuum pressure-swing adsorption (VPSA) is an alternative carbon dioxide separation system considered for TGR-BF. Pressure-swing adsorption uses adsorbents with strong sorption affinity for carbon dioxide such as Zeolite 13X, metal–

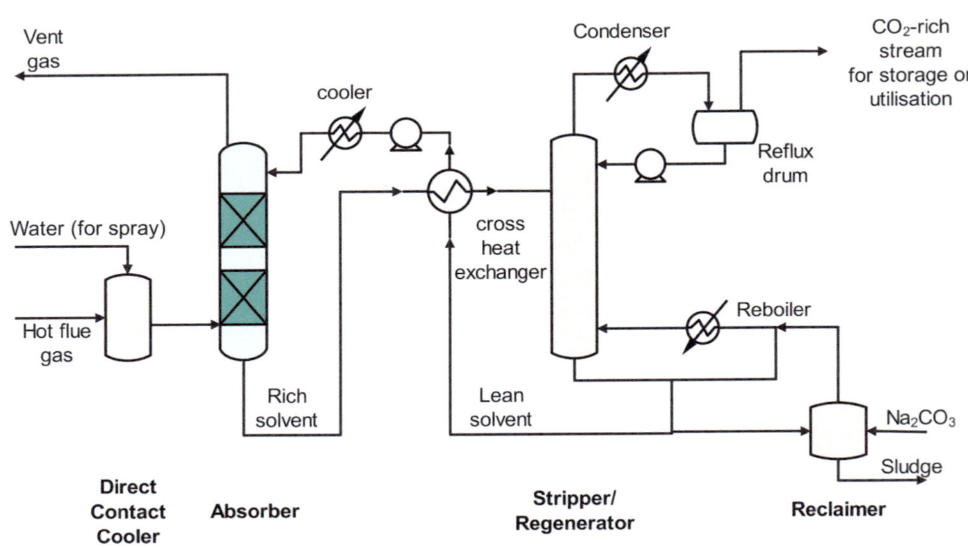

Fig. 3 Absorption-based carbon dioxide capture

organic frameworks and silica gel. Typically, VPSA is a cyclic process with five steps—(a) pressurisation, (b) adsorption, (c) depressurisation, (d) evacuation under vacuum and (e) purge. A comparison of absorption and VPSA-based CO_2 separation technologies [11] suggests that the energy consumption for VSPA (1.05 GJ/tCO_2) is substantially lower than that of an amine-based absorption process (3.81 GJ/tCO_2). The quality of energy requirements is different for both processes. Energy demand for regeneration in an amine process is substantially high at 3.2 GJ/tCO_2. On the other hand, VPSA consumes marginally higher electricity (0.29 MWh/tCO_2) compared to an amine process (0.17 MWh/tCO_2). Given that the VPSA has lower energy demand, the cost of CO_2 separation by this process is estimated to be 25–40% lower than amine-based absorption process. However, these estimates are price sensitive and availability of waste heat for regeneration improves the economics for absorption process. Efforts are also underway to bring down the regeneration energy demand to 1.8 GJ/tCO_2 for CO_2 absorption process [11].

5 Mineral Carbonation Using Iron and Steel Slags

Mineral carbonation is a process of converting gaseous carbon dioxide into sparingly soluble minerals such as calcium carbonate. Typically, naturally occurring minerals or industrial alkaline wastes that are rich in calcium or magnesium are used as a feedstock.

5.1 Use of Iron and Steel Slags

Iron and steel slags are generated as a waste or as a low-value byproduct during the steelmaking process. Nearly 0.5 tonnes of slags are produced per tonne of crude steel. Steel slags have a high calcium content (Ca $\geq$ 30 wt%) and several studies found them to be a suitable feedstock for mineral carbonation [16, 17]. Typical mineral phases in a steel slag include dicalcium silicate, brownmillerite, RO phases containing free CaO, MgO and solid solution of MgO–FeO–MnO [16]. As such, steel slags find low utilisation and considered as a low-value byproduct, and very often ends up in landfills. Typically, slags are employed in construction activities as aggregate (e.g. as a replacement for natural gravel) or supplementary cementitious material (as a replacement for Portland cement). In case of steel slags, excess free lime (CaO) upon hydration expands. This volume instability is a significant concern for its direct use in construction industry. Prior mineral carbonation of steel slags, where free lime is converted into calcium carbonate, is known to alleviate this problem and allows gainful utilisation in the construction industry.

5.2 Mineral Carbonation Routes

As such, Gibbs free energy change (ΔG_r°) for mineral carbonation reaction is negative at standard atmospheric conditions and hence a thermodynamically spontaneous process. An example mineral carbonation reaction using dicalcium silicate is given below:

$$0.5Ca_2SiO_4 + CO_2(g) -> CaCO_3(s) + 0.5\ SiO_2(s) \qquad (2)$$

$$\Delta H_r^\circ = -116.2\ \text{kJ/mol};\ \Delta G_r^\circ = -90.1\ \text{kJ/mol}$$

This reaction is also mildly exothermic and does not need any additional heat input. Additionally, mineral carbonates such as $CaCO_3$ show low solubility in water and do not undergo thermal decomposition even at temperatures as high as 300 °C and are stable for thousands of years [18]. Despite these thermodynamic advantages, owing to multiple physicochemical characteristics, mineral carbonation rates achieved in a gas–solid (CO_2-mineral) reactor at room temperature were too low to make a commercial process. In the last three decades, several research efforts to improve the kinetics led to multiple novel carbonation routes, which were reviewed and assessed elsewhere [17].

The more promising mineral carbonation routes are based on reactions in aqueous media. Currently, mineral carbonation in aqueous media is understood to be a three-step process—(a) absorption of CO_2 into aqueous media, (b) leaching of Ca^{2+} into solution and (c) precipitation of calcium carbonation from the solution. Depending on the number of process steps, these routes are categorised into direct and indirect carbonation. Direct carbonation routes are typically single step processes with a three-phase reactor where gaseous CO_2 is absorbed into an aqueous slurry containing mineral particles. Depending on the liquid-to-solid ratio, these routes are further categorised into thin-film carbonation and slurry carbonation [19]. Slurry-based carbonation (CO_2-H_2O-mineral) typically uses a liquid-to-solid (LS) ratio greater than 2 L/kg, while the thin-film carbonation is performed at a lower LS ratio. On the other hand, indirect carbonation routes are multi-step processes where carbon dioxide absorption and (or) mineral precipitation steps are achieved in separate unit operations. Single-step three-phase (CO_2-H_2O-slag) carbonation methods are inflexible, and their performance is limited by lack of harmony in the rates of absorption, dissolution and precipitation. For instance, absorption and precipitation are favoured in alkaline media while leaching of Ca^{2+} is conducive only in acidic media. Thus, multi-step indirect routes are a preferred option.

Early research efforts in the development of indirect carbonation were focused on pH-swing process (Fig. 4a).

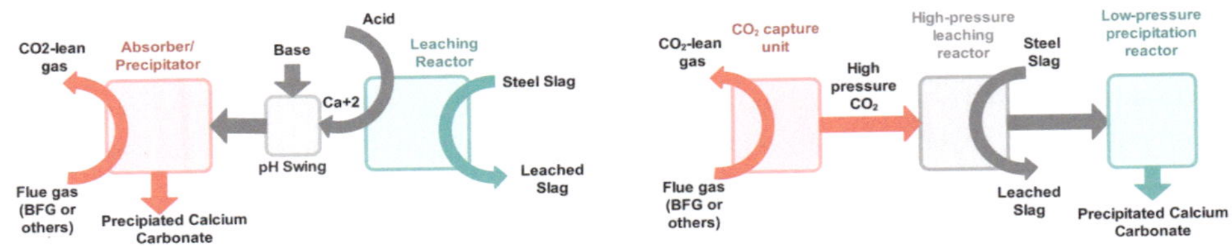

(a) pH-swing

(b) Pressure-swing

Fig. 4 Indirect mineral carbonation routes to produce precipitated calcium carbonate

pH-swing is a three-step process where acid-mediated Ca^{2+} leaching was followed by a series of pH-swing and absorption steps. This multi-step process can produce precipitated calcium carbonate (PCC), a valuable specialty chemical used as a filler in paper and plastic industry. However, regeneration of acid and base makes the process energetically and economically unviable. An alternative method, which avoids the use of mineral acids, is pressure-swing carbonation shown in Fig. 4b. Here, the dependence of calcium carbonate solubility on CO_2 pressure is exploited to recover PCC from steel slag [20]. In the dissolution reactor, high-pressure carbon dioxide is absorbed into the aqueous slurry. High pressure carbon dioxide keeps the solution pH low (3.5–5.5) to accelerate the calcium leaching into the solution. Moreover, calcium carbonate solubility rises with CO_2 pressure to keep Ca^{2+} ions in the solution. The Ca-rich aqueous solution is separated from residual slag by filtration and depressurised to atmospheric conditions. Pressure-swing step desorbs the carbon dioxide gas, and consequently increases the pH and lowers calcium carbonate solubility. Thus, calcium carbonate precipitate is spontaneously discharged from the solution [20]. The main limitations are in obtaining high-pressure carbon dioxide at low cost from a CO_2 capture unit and recycling the liberated carbon dioxide during the pressure-swing step.

Recent work by Ragipani and co-workers [21] investigated mineral carbonation in alkali bicarbonate/carbonate solutions. Experiments with class C coal fly ash showed very promising results and reported a maximum carbonation efficiency of 95%. Carbonation efficiency is calculated as the % of total calcium converted to calcium carbonate. As shown in the following reaction 3, mineral carbonation in alkali carbonate solutions generate hydroxide solution as a byproduct. Residual slag containing calcium carbonate, unreacted phases and secondary precipitates such as silica can be separated from the aqueous solution by filtration.

$$0.5\,Ca_2SiO_4(s) + Na_2CO_3(aq) + H_2O \longrightarrow CaCO_3(s) + 0.5SiO_2(s) + 2NaOH(aq) \qquad (3)$$

$$\Delta H_r^\circ = -6.8\ \text{kJ/mol}$$

This process has the advantage that the reaction is thermodynamically spontaneous at atmospheric conditions and generates a strong alkali hydroxide solution that can readily absorb carbon dioxide and regenerate the solvent.

6 Coupling CO_2 Capture and Mineral Carbonation in Alkaline Media

Recent work on mineral carbonation process in alkali carbonates led to a coupled mineral carbonation and CO_2 capture process. As shown in Fig. 5, hydroxide-bicarbonate/carbonate solution is recycled between carbon dioxide capture unit and mineral carbonation reactor. In the CO_2 capture unit, carbon dioxide-rich gaseous stream, such as BF top gas, is contacted with strongly alkaline hydroxide solution (typically, at a pH as high as 13.5–13.9). Absorption of CO_2 generates bicarbonate/carbonate solution with a pH as low as 8.0. In the mineral carbonation reactor, Ca-rich steel slag reacts with the solution to form calcium carbonate and regenerate the hydroxide solvent, which is recycled to CO_2 capture unit. Thus, a coupled two-step process is achieved with recyclable solvent.

$$CO_2(g) + 2\,NaOH(aq) \longrightarrow Na_2CO_3(aq) + H_2O \quad \Delta H_r^\circ = -109.4\ \text{kJ/mol} \qquad (4)$$

Alkali hydroxides and carbonates are well studied solvents for carbon dioxide absorption [22–25]. In the more recent past, alkali hydroxide-based solvents are also considered for direct removal of carbon dioxide from air [21, 26]. In CO_2 capture systems, the proposed process can be an alternative to amine-based absorption processes, especially in steel industries where Ca-rich slag residues are available for solvent regeneration via mineral carbonation.

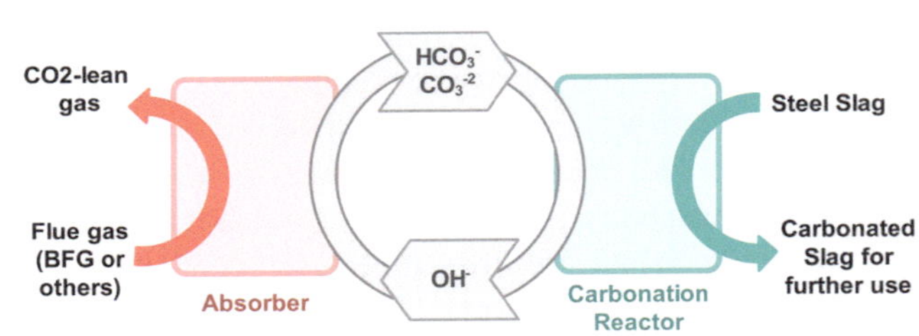

Fig. 5 Carbon capture coupled with mineral carbonation

As stated in Sect. 4, in a conventional amine-based CO_2 capture process, solvent regeneration is performed by temperature-swing. This requires heat input to the stripper (up to 3.2 GJ/tCO_2). On the other hand, in the proposed process, solvent is regenerated by mineral carbonation reaction. Given that the mineral carbonation is spontaneous at atmospheric conditions, there is no additional heat input needed. Thus, the proposed process allows one to use very strong solvents such as alkali hydroxides without any energy penalty on solvent regeneration. Moreover, carbon dioxide is obtained as stable carbonates and does not require compression and sequestration efforts. Thus, the energy requirements and hence the overall cost of avoiding CO_2 using the proposed mineral carbonation process is expected to be lower than that of amine-based absorption and VPSA processes.

7 Challenges and Opportunities

Coupling carbon dioxide capture with mineral carbonation is a promising CO_2 abatement technology. The serendipitous availability of Ca-rich slags and coal ashes is an immense opportunity for iron and steel mills. Currently, mineral carbonation-based technologies are still under lab-scale development and far from commercialisation. On field pilot-scale studies followed by a detailed technoeconomic and life cycle assessments are needed to derisk the process for commercialisation. Other challenges include sustained availability of Ca-rich wastes and minerals and identifying ways to utilise the carbonated slags.

Competing Interests The authors have no conflicts of interest to declare that are relevant to the content of this chapter.

References

1. Fisch-Romito V (2021) Global Environ Change 70. https://doi.org/10.1016/j.gloenvcha.2021.102362
2. IEA (2021) Iron and steel technology roadmap towards more sustainable steelmaking part of the energy technology perspectives series
3. Bains P, Psarras P, Wilcox J (2017) Co2 capture from the industry sector. https://doi.org/10.1016/j.pecs.2017.07.001
4. MOS (2023) Annual report
5. Walsh B, Ciais P, Janssens IA, Peñuelas J, Riahi K, Rydzak F, Vuuren DPV, Obersteiner M (2017) Nat Commun 8. https://doi.org/10.1038/ncomms14856
6. Assen NVD, Voll P, Peters M, Bardow A (2014) Chem Soc Rev 43:7982. https://doi.org/10.1039/c3cs60373c
7. Fruehan RJ, Fortini O, Paxton HW, Brindle R (2000) Theoretical minimum energies to produce steel for selected conditions
8. Sahoo M, Sarkar S, Das AC, Roy GG, Sen PK (2019) Steel Res Int 90. https://doi.org/10.1002/srin.201900034
9. Htet TT, Yan Z, Spooner S, Degirmenci V, Meijer K, Li Z (2021) Fuel 289. https://doi.org/10.1016/j.fuel.2020.119890
10. Zhang X, Jiao K, Zhang J, Guo Z (2021) A review on low carbon emissions projects of steel industry in the world. https://doi.org/10.1016/j.jclepro.2021.127259
11. Birat JP (2010) Carbon dioxide (CO2) capture and storage technology in the iron and steel industry. Elsevier, pp 492–521. https://doi.org/10.1533/9781845699574.5.492
12. Rainer R, Miguel AAM, Serge R, Luis DS (2013) Best available techniques (bat) reference document for iron and steel production
13. Higashi I (1982) Trans ISIJ 22:57
14. Kuramochi T, Ramírez A, Turkenburg W, Faaij A, (Elsevier Ltd, 2011), vol 4, pp 1981–1988. https://doi.org/10.1016/j.egypro.2011.02.079
15. Tobiesen FA, Svendsen HF, Mejdell T (2007) Ind Eng Chem Res 46:7811. https://doi.org/10.1021/ie061556j
16. Ragipani R, Bhattacharya S, Akkihebbal S (2020) Waste Manag 117. https://doi.org/10.1016/j.wasman.2020.08.008
17. Ragipani R, Bhattacharya S, Suresh A (2021) React Chem Eng 6:1152. https://doi.org/10.1039/D1RE00035G. http://xlink.rsc.org/?DOI=D1RE00035G
18. Lackner KS (2003) Science 300:1677. https://doi.org/10.1126/science.1083529
19. Schnabel K, Brück F, Pohl S, Weigand H (2022) Waste and Biomass Valorization 13:4319. https://doi.org/10.1007/s12649-022-01784-z
20. Ragipani R, Bhattacharya S, Suresh A (2019) Proc R Soc A: Math Phys Eng Sci 475. https://doi.org/10.1098/rspa.2018.0830
21. Ragipani R, Sreenivasan K, Anex R, Zhai H, Wang B (2022) ACS Sustain Chem Eng 10. https://doi.org/10.1021/acssuschemeng.1c07867
22. Astarita G (1963) Ind Eng Chem Fundam 294–297. https://pubs.acs.org/sharingguidelines
23. Holmes G, Keith DW (2012) Philos Trans R Soc A: Math Phys Eng Sci 370:4380. https://doi.org/10.1098/rsta.2012.0137
24. Masaki BK (1931) J Biochem XIII:211
25. Danckwerts P, Kennedy A (1958) Chem Eng Sci 8(3):201. https://doi.org/10.1016/0009-2509(58)85027-7. https://www.sciencedirect.com/science/article/pii/0009250958850277
26. Keith DW, Holmes G, Angelo DS, Heidel K (2018) Joule 2:1573. https://doi.org/10.1016/j.joule.2018.05.006

Decarbonizing India's Steel Sector: Green Steel—in Need of Timely and Far-Sighted Policies to Attract Investment

Mousami Prasad and Anisha Ahuja

1 Introduction

Typically manufacturing of steel results in the emission of significant amounts of carbon dioxide. These emissions are a major contributor to global warming causing significant changes in the climatic conditions across the world. The sector is responsible for about 10% of the global carbon emissions with an average emission intensity of 2.1 tCO_2/t of steel [1]. Primary steelmaking through the blast furnace route is a very common and the most emission-intensive process where iron ore and metallurgical coke, a source of carbon, are heated to very high temperatures to produce pig iron which is then converted into steel in either open-hearth furnaces or oxygen-based converters. This two-step integrated steelmaking process is responsible for producing 75% of the global crude steel.

India is the world's second largest crude steel producer after China, with 6.6% of global share at 125 Mt (million tons) in 2022. The sector contributes 2% to the national economic output measured in Gross Domestic Product (GDP) and generates employment for nearly 2.5 million people directly and indirectly [4]. However, India's steel sector is highly energy-intensive and emission-intensive compared to the global average. Producing 1 ton of crude steel in India consumes ~6 GJ of extra energy, resulting in additional emissions of ~350 kg CO_2 than the global average. Domestically, the sector is responsible for ~10% of CO_2 emissions. With the rising demand for steel, its production is projected to at least double by 2030 from the 2019 levels [2, 4], and if there is no change in the way of producing steel, the emissions too will double.

While incremental decarbonization measures, like energy efficiency, electrification, increasing the use of scrap, and a switch to renewable energy, can reduce the energy demand and CO_2 emissions, it will be insufficient to put the sector in line with India's climate commitments and decarbonization targets of the steel sector at the global level, calling for fundamental shifts in the technology and process in the steelmaking [2]. Currently, the technologies that can replace coke in primary steelmaking are in the pilot stages. One of the most promising technologies emerging is hydrogen-based steelmaking, in which green hydrogen, which is generated using renewable energy sources like solar and wind, is used to replace coke used in the blast furnace. The resulting steel reduces emissions by ~99% and is referred to as fossil-free or zero-carbon steel (a misnomer, as steel is an alloy of Fe and C) or green steel. The last one is the most apt term.

Hydrogen-based steelmaking (i.e., green steel) is projected to produce 13–30% of the crude steel in India by 2030 and 40–70% by 2050 and is likely to be aligned with the global net-zero goals by 2050. However, the cost of producing green steel is currently high, and there are uncertainties about the cost of hydrogen. Under national hydrogen missions, India aims to produce green hydrogen for domestic consumption and export. The use of hydrogen in steel has been included as one of the application areas under the pilot projects [3]. However, these plans lack detailed guidelines, particularly for steel, and the on-ground rollout has not yet happened necessitating a more focused policy on green steel. Such a policy is needed for long-term directionality and reducing risk for the private sector investment.

In the absence of a timely policy, the steel companies will continue to lock in their investments in conventional technologies using fossil fuel energy. While few companies may invest in green steel technology, but transformation at the sector level is unlikely. Forty-two percent of the nation's steel

M. Prasad (✉) · A. Ahuja
Department of Management Sciences, Indian Institute of Technology, Kanpur, India
e-mail: mousami@iitk.ac.in

M. Prasad
Institute for Climate, Energy & Disaster Solutions, Australian National University, Canberra, Australia

S. Narayanan et al. (eds.), *Materials, Devices and Systems for Sustainability*, IITK Directions 8,
https://doi.org/10.1007/978-981-96-7295-0_9

is produced from the BF-BOF route which uses coking coal. Most of coking coal is imported from countries like Indonesia, Australia, South Africa, and Russia and as these nations move away from fossil fuel exports, India's steel production targets can be strained. The other fallout might be that the sector's exports become uncompetitive in the global markets considering efforts from Europe and other nations favoring low-emission or green steel. While there may be push back from countries on such measures, it is vital and timely to rethink about India's steel sector. Investment in the development of scalable and replicable clean technology solutions can reduce fossil fuel imports, cut down carbon emissions and create new growth opportunities. Low emission steel could be the growth engine for the country.

The contents of the paper are organized as follows. Section 2 discusses the role of green steel in India followed by the projections of its production in the years 2030 and 2050 in Sect. 3. Section 4 discusses the challenges and uncertainties and the way forward. Section 5 provides a summary and concluding remarks.

2 Why Does Green Steel in India Have a Role?

India's steel sector is responsible for ~9% of national energy consumption and ~10% of national CO_2 emissions (excluding the indirect emissions from power generation and imported heat) [4] (Fig. 1). To produce 1 ton of crude steel in India, about 26 GJ of energy is used. About 2.3 tCO_2 is emitted which is ~20–25% higher than the global average of 20 GJ of energy consumption per ton of crude steel and emission of 1.9 tCO_2 per ton of crude steel [2, 5].

Globally, there are two main steelmaking routes: the blast furnace-basic oxygen furnace (BF-BOF) route and the electric arc furnace (EAF) route. The conventional steelmaking method—BF-BOF uses iron ore, coal, and recycled steel, while the EAF route uses recycled steel and electricity. The key difference between the routes is the type of raw material used, which makes the two methods different in terms of energy consumption and emissions [1]. The current emission intensity levels do not align with global, regional, and sectoral climate targets. Many nations including India have made net-zero commitments that will require emissions from steel to reduce significantly needing fundamental technology and process changes, including green steel technologies [2, 4, 5, 7]. IEA estimates that between the year 2020 and the year 2050, half of the cumulative CO_2 emissions from the Indian steelmaking can be lowered by material efficiency combined with steel companies moving to the best available technology [2]. However, gains from material efficiency are very uncertain given that it depends upon the measures taken by the end-user sectors of steel and is difficult to predict. Another alternative is to increase the use of scrap in steelmaking. However, scrap steelmaking alone may not be sufficient for decarbonizing the Indian steel industry. Looking at the increasing demand for steel in India with limited availability of scrap steel combined with insufficient infrastructure for collection and processing scrap and the need for high-quality steel scrap, India needs a comprehensive approach [8]. While scrap steelmaking can play a role, integrating breakthrough technologies and strategies is essential to achieve net-zero targets. Hence, low-emission technologies to replace fossil fuels in steelmaking such as with electricity from renewables, biomass or hydrogen can play transformational role in reducing emissions. The resulting steel is often referred to as green steel or fossil-free steel and can reduce emissions up to 99% (IPCC AR6). After years of research in low-carbon technologies, a few breakthrough technologies like Molten Oxide Electrolysis (MOE) using electricity to directly reduce iron oxide to produce steel in a single step, hydrogen-based direct reduced iron (H_2-DRI), and Carbon Capture, Utilization, and Storage (CCUS)/Carbon Capture and Storage (CCS) are finally coming to the commercial stage, making green or low-carbon steel a reality. Another alternative involving hydrogen plasma makes a viable alternative for hydrogen-based steel. It allows the production of liquid iron in a single step by creating a hydrogen plasma arc. The by-product of this reduction process is water, which makes steelmaking carbon-free [12].

However, there is no consensus on what constitutes green steel amongst countries and companies. Some studies describe green steel in the technology terms such as zero-carbon steel [9], fossil-free steel [10], and hydrogen-based steelmaking [11]. Some companies use emissions as a criterion to define green steel and include any low-emission technology, carbon capture, recycling, increased use of scrap, use of zero-carbon fuels like hydrogen, and electricity from renewable energy sources.

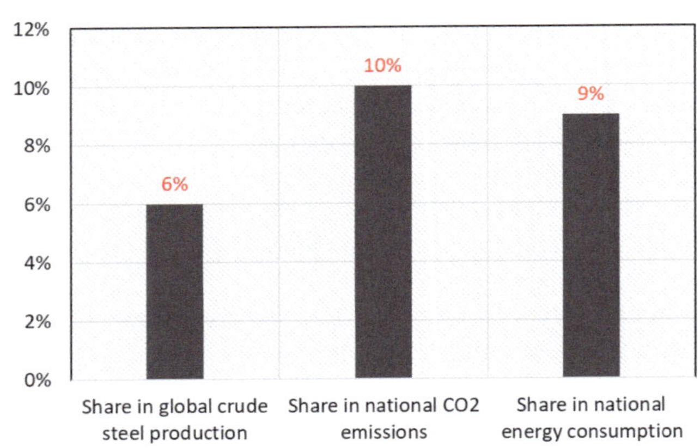

Fig. 1 Indian steel industry profile. *Source* Authors compilation based on 2019 data from IEA 2019, [6]

Table 1 What is green steel? Explanations from selected organizations

Source	Description
AFRY [15]	The Green steel manufacturing process focuses on using electricity and non-fossil fuels, consequently reducing greenhouse gas emissions compared to conventional steel production
Leadership Group from Energy Transition [16]	Low-carbon initiatives within the primary steel sector, molten oxide electrolysis (MOE), electrowinning and electric smelting reduction (ESR), and hydrogen direct reduction (H-DRI) projects when using renewable sourced hydrogen—commonly known as green hydrogen—for the reduction of iron ore, certain carbon capture, and storage projects with timeline
H2 Green Steel [17]	Electricity from fossil-free sources and green hydrogen instead of coal, giga-scale electrolysis, DR reactor, electric arc furnace, continuous casting, and downstream finishing
SSAB [18]	Distinguishes between green steel and fossil-free steel HYBRIT* process involves no fossil fuels or CO_2 emissions

* HYBRIT: Hydrogen Breakthrough Ironmaking Technology
Source Authors compilation based on organizations' websites and reports

For instance, Liberty Steel UK GreenSteel initiative includes recycling, and upcycling of the scrap, the use of renewable energy to power the electric arc furnace [13]. In India, Kalyani Group launched the country's first green steel brand "Kalyani FeRRESTA" in December 2022, produced through an electric arc furnace using renewable energy generated electricity with an emission intensity of less than 0.19 tCO_2e/t of crude steel [14]. A summary of various definitions and explanations of green steel by different bodies is given in Table 1.

For the present study, we define green steel as a primary steelmaking process that does not use fossil fuel energy sources. These technologies include using hydrogen-based direct reduced iron (H_2-DRI) in which hydrogen is used to reduce iron ore to iron and then using electric arc furnace to produce steel or aqueous electrolysis route which uses an electrolyte solution and electric current followed by electric arc furnace; or molten oxide electrolysis using electricity to directly reduce iron oxide to produce steel in a single step. Although at pilot stages, H_2-DRI steelmaking is emerging as a promising solution with projects already at different stages of progress as per Green Steel Tracker [16] like Steel4Future at ArcelorMittal focused on large scale industrial plant for DRI by 2026, HYBRIT at SSAB aimed at bringing steel produced by 2026, H2V CAP at CAP SA a project of a steel producer from Chile aiming to run pilot phase until 2025 to generate 25,000 tons/year of sponge iron, HYFOR at Voestalpine.

In addition to the climate argument there are at least two other perspectives on why green steel could be an advantage for India. First, the role green steel can play in energy security. India currently produces 42% of its steel using integrated steelmaking (blast furnace–basic oxygen furnace route), which uses coking coal or metallurgical coal (coal used in the metal refining process). Typically, 800 kg of coking coal is needed to produce 1 ton of crude steel. However, India does not have sufficient coking coal as Indian coal has a higher content of moisture with lower calorific value and hence relies on imports. For this, India imports about 50 Mt of coking coal each year, which is about 25% of India's total coal imports and accounts for 40% of the coal import value [7]. Apart from this, non-coking coal and natural gas are also imported. Therefore, any transition away from fossil fuel energy sources to domestically generated energy sources will have direct impact

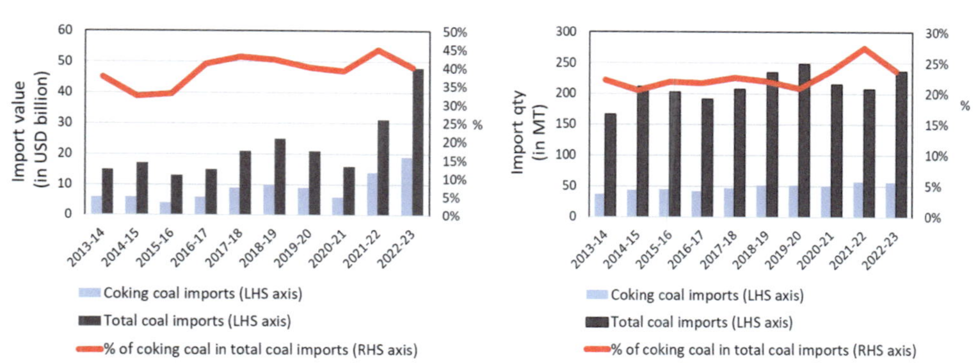

Fig. 2 India's coking coal imports in value and in quantity for the last 10 years [7] *Source* Authors summarization based on data from GoI, Ministry of Coal

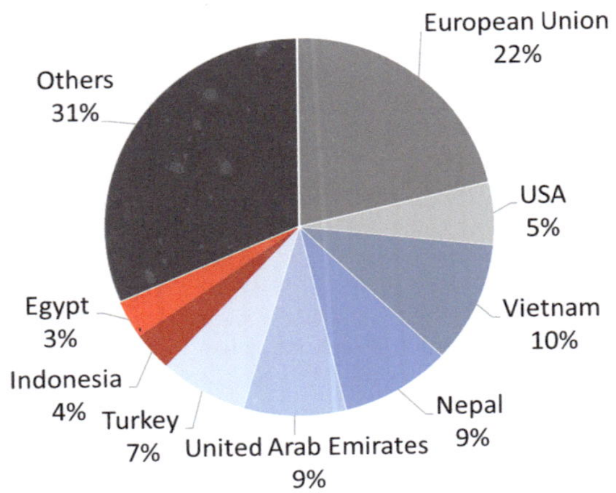

Fig. 3 India's steel exports to selected countries in the year 2022 [20]
Source Authors compilation based on data from US DCEE

on energy security and lower energy import bills, freeing up foreign reserves for other uses (Fig. 2).

Second, the green steel industry can spur economic growth and create green jobs. Europe's introduction of a carbon border adjustment mechanism (CBAM) for carbon-intensive imports into the region has created a demand center for green products such as green steel. Currently, India exports 11 Mt or 8.8% of the steel valued at USD 12.7 billion [19], of which 20% is exported to European Union states such as Italy, Belgium, and Spain (Fig. 3). India can capitalize on this export market potential and probably make it bigger through a high-value green steel proposition. From a net importer of fossil fuel energy, India can become an exporter of renewable energy-based products like green steel.

Domestically, the iron ore-rich regions in India are co-located with the coal mining regions in the eastern side (Fig. 4) such as states of Jharkhand, West Bengal, and Orissa which can exploit their co-location advantage to produce green iron / green steel. This can offset the likely decline in revenues from coal production by new opportunities in green steel. This will have huge implications for domestic regional economic growth by creating a new economic activity and jobs.

Further, if we map India's solar potential with that of iron ore mining areas, we find an overlap between the two resources, suggesting a natural co-location advantage (Fig. 5). A recently discovered iron ore mine in the western state of India with the potential of 2 billion tons of iron ore deposit [21] could be a major potential given that Rajasthan has the highest solar potential in the country and solar parks which can provide zero-carbon electricity to process the iron ore. Similarly, the existing assets in the integrated steelmaking route or BF-BOF are situated in regions with solar potential and/or are well connected to the national grid (Fig. 6).

3 Projection for India's Production of Green Steel by the Year 2030 and the Year 2050

Steel consumption per person per year in India at 76 kg is almost one-third of the global average at 232 kg and much lower than some of the developed nations such as the United States at 290 kg and Germany at 426 kg [22] which is likely to increase as the Indian economy grows. National Steel Policy 2017 intends to increase the steel demand by enhancing per capita steel consumption to 160 kg in India.

As per IMF forecasts Indian economy is expected to grow above 6%, which is twice the global growth rate [19] and this will increase the steel demand. Currently, 40% of steel is used in the construction industry that includes capital projects and infrastructure, followed by 22% in automobile and transportation and another 10% in consumer goods [23]. Each of these sectors is experiencing growth, supported by sector-specific government policies such as PM GatiShakti National Master Plan and Make in India initiatives that will increase potential use of steel and steel products including engagement from Railways, Defense, Petroleum, Natural Gas, Housing, Aviation, and Road Transport and Highways.

Various estimates on steel production also reflect this increased demand in steel production which is projected to reach 191–255 Mt by the year 2030, and 339–570 Mt by the year 2050 (Fig. 7). This is an increase of 66–130 Mt in steel production in the 8 years of 2022–2030 or an increase of 8–16 Mt every year of steel production until 2030.

To meet these production levels, energy consumption is expected to increase to 1.8 times by 2030 and 3.5 times by 2050 from the 2019 levels. Given the current energy mix of the sector, this will translate into higher CO_2 emissions. Business, as usual, therefore is not an option if India is to meet its net-zero commitments. To be in line with global net-zero targets, IEA estimates that the emissions from steelmaking should be 350 Mt CO_2 by the year 2030 at a production level of 200 Mt implying that the emission intensity of steelmaking should reach 1.6–1.8 tCO_2/t steel in the year 2030 which is a reduction of 20–30% from the emission intensity levels of the year 2019 and currently the global best levels. Getting the emission intensity to these levels will need serious mitigation efforts.

With India's current steelmaking technology, timely action and investment by the companies can help the sector decarbonize. In 2019, India produced 26% of steel through the Direct Reduced Iron (DRI) route either using coal or gas, which is much higher than the global level of 7% (Fig. 8). For the transition to hydrogen, the DRI plants could be the first test bed and some Indian companies are already showing interest in trialing such as JSW Steel's announcement of using hydrogen in its Vijayanagar DRI Plant [25].

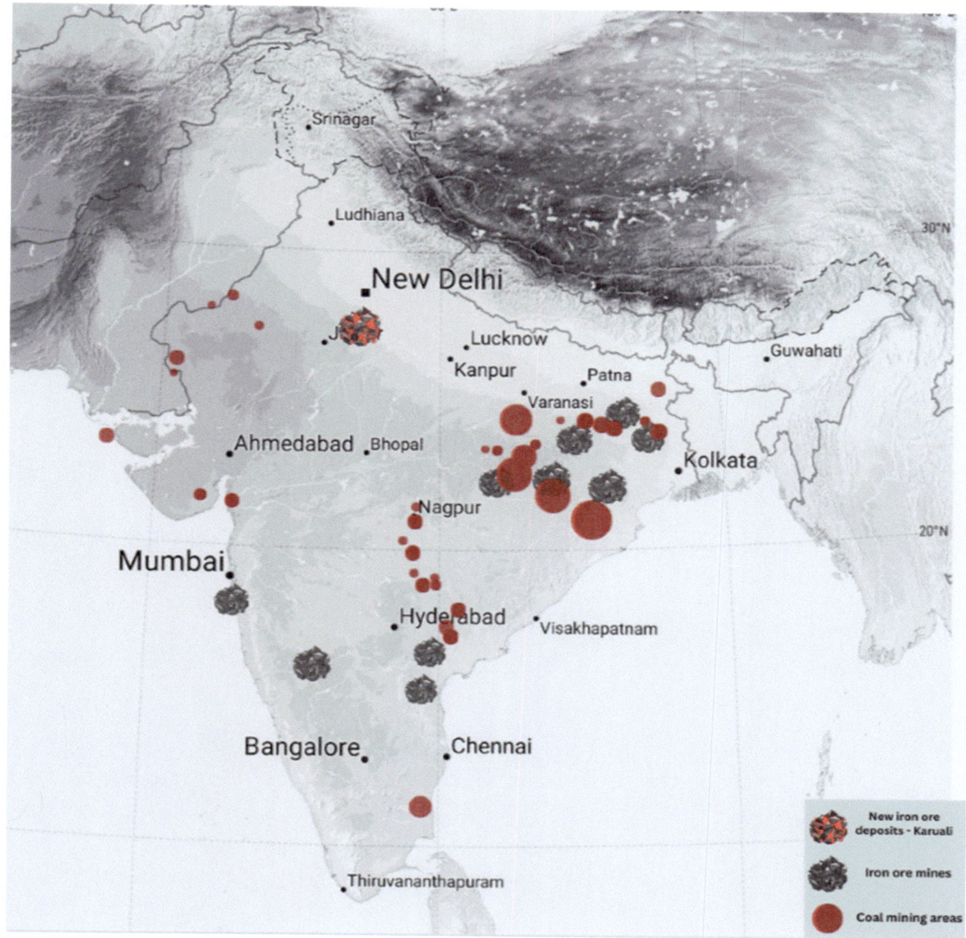

Fig. 4 Iron ore mines in India mapped with coal mines *Source* Author compilation from Global Energy Monitor and Ministry of Steel

Various studies on projecting steel technology routes in India by the years 2030 and 2050 suggest a role for hydrogen-based steelmaking and estimate that the share of H_2-DRI in total steelmaking could range from 13 to 30% of steel by 2030 and 16 to 67% by 2050 to be in line with net zero by 2050 (Fig. 9). Some studies predict that H_2-DRI technology could produce more than half of the steel by 2050 (Fig. 9).

Existing studies also suggest there will be more green steel manufacturing technologies such as molten oxide electrolysis and electrowinning in addition to hydrogen-based steelmaking (Fig. 9).

It is also clear that in the future, steel will be produced in many ways than it is currently. For instance, HIsarna technology by TATA Steel is an alternative to blast furnaces. The raw materials could be used in powder form and can directly be converted into liquid iron, making steel with 20% less carbon emissions [26]. However, estimates on the share of conventional blast furnaces and basic oxygen furnaces in the future vary widely among the studies. Certain studies project that the share of BF-BOF is likely to come down from 42 to 22% in 2030 and completely be phased out by 2050. The other set of studies, taking a more conservative approach, expects the BF-BOF share to increase up to 56% by 2030 and remain at about 50% by 2050.

A few studies also note that existing technology will be all retrofitted with carbon capture storage by 2050 (Fig. 9). Each of the pathways has its financial implications in terms of investments and return on investments, which we have not examined in this study but will be necessary for understanding the economic feasibility.

4 Mainstreaming Green Steel Technology: Challenges and Uncertainties

Decarbonizing steel production has many pathways, and it is apparent that it will not be one single technology but multiple technologies. It is also clear that carbon capture and storage/utilization will be used alongside zero-carbon energy

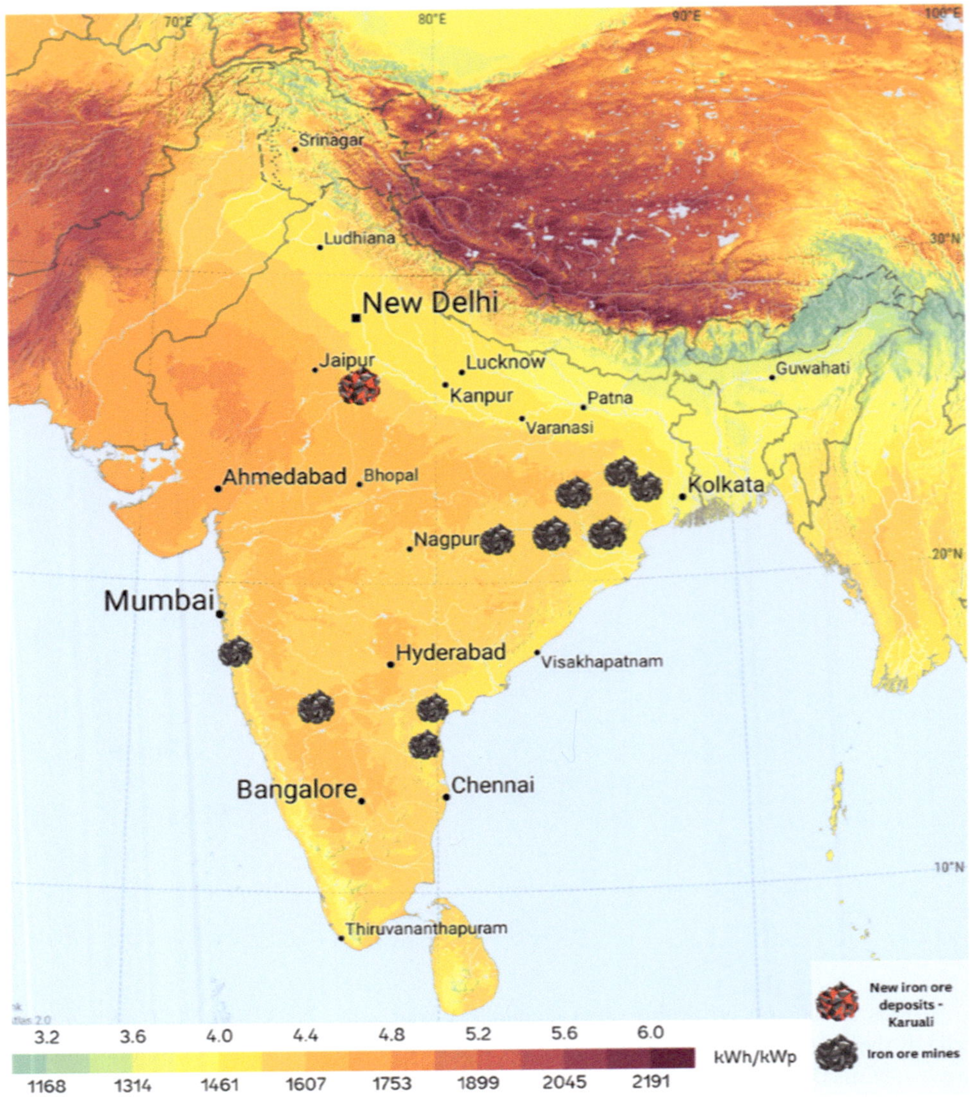

Fig. 5 Iron ore mines and new deposits in Karauli district, Rajasthan, mapped with solar irradiance map *Source* Author compilation from Global Energy Monitor, Ministry of Steel and Global Solar Atlas

generation technologies. However, each pathway has its timeline emission reduction potential and will need investments to be committed now. A typical conventional Indian BF-BOF plant of 3 Mt steel per annum needs an investment of INR 255 billion (USD 3 billion) and a construction time of 5 years [3] and the requirements could be much higher for the green steel.

Some studies have also argued for gas-based hydrogen or blue hydrogen as a transition or intermediate solution. Blue hydrogen technology uses natural gas for steelmaking through steam methane reforming (SMR), which releases carbon dioxide. Generating hydrogen from natural gas may result in higher GHG emissions (prominently carbon dioxide and methane) than directly burning natural gas [27]. Emissions from blue hydrogen can be lower if there is CCS with a 90% capture rate but CCS is costly, and the capture rates in reality are much lower. Hence, adopting blue hydrogen will provide no benefit and hinder progress toward achieving net-zero targets in the steel sector. [28]. For India, blue hydrogen can be an additional burden given that most of the gas is imported which can strain India's energy imports and raise energy security concerns.

To produce 1 ton of green steel, about 60 kg of H_2 is required [29]. Current estimates of producing green hydrogen in India are about \$4–7 per kg of H_2 which is expected to reduce to \$1.50/kg by 2025 and to \$1/kg by 2030 [10]. This does not include the cost of transport and storage. To achieve the target prices of hydrogen production, the cost of electricity from renewable energy must come down. 1 kg of hydrogen needs about 50–60 kWh of electricity. This requires massive investments in the scaling of generation from renewable energy and electrolyzer capacity for hydrogen production. The role of energy storage will also be critical.

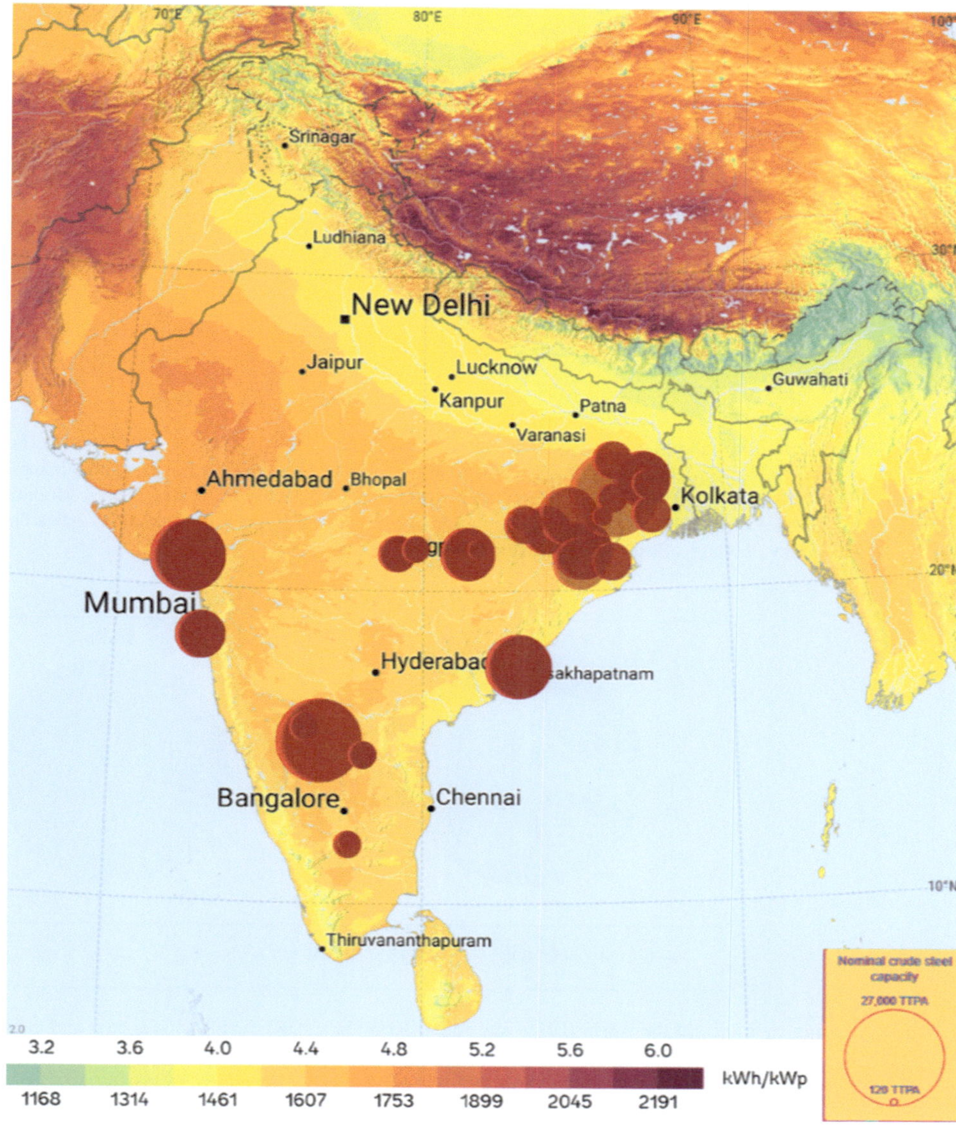

Fig. 6 Iron and steelmaking plants (BF-BOF; BF-DRI) mapped with solar irradiance map in India *Source* Author compilation from Global Energy Monitor, Ministry of Steel and Global Solar Atlas

Further to produce, say 25 Mt of green steel from the hydrogen route by 2030, which is in line with projections for India to achieve the net-zero target by 2070 [30], will need 1.5 Mt (25 Mt × 60 kg) hydrogen. To be able to produce 1.5 Mt of hydrogen, 75–90 TWh (1.5 Mt × 50–60 kWh) of electricity will be needed by 2030 which is equivalent to 8–10% of total electricity generated in India or the entire electricity produced from solar PV in India in the year 2022. To meet this additional demand for renewable electricity, huge investments will be required in upscaling renewable energy capacity and the production of electrolyzers to produce hydrogen. To get a sense of the electrolyzer capacity needed to produce 1.5 MtH_2, the electrolyzer capacity of 2.1 Mt H_2 will be needed assuming an electrolyzer efficiency of 70%. Considering an electrolyzer size of 1 MW which can produce 17–20 kg of hydrogen per hour or 400–480 kg H_2 per day assuming the continuous supply of renewable electricity translates into 146–173 tH_2/year assuming 360 days of operation. To produce this amount of hydorgen, it will need 8,680–10,212 electrolyzers by 2030. Similarly, considering the typical DRI plant size of 1 Mt per annum in India, about 20–25 green steel plants will be needed by 2030 to produce 25 Mt green steel.

Cost estimates from pilot initiative of H2-DRI EAF that produced green steel in 2020 in HYBRIT plant in Sweden [31]. Suggests that the costs can vary widely depending on the cost of hydrogen production and the proportion of steel scrap which could range from 10–100% of the costs of steel from blast furnaces [10]. Uncertainty in the costs has kept companies from committing to new H_2-DRI plants. Another concern is the demand for green steel. It is understood that the cost of green steel will be higher than conventional steel

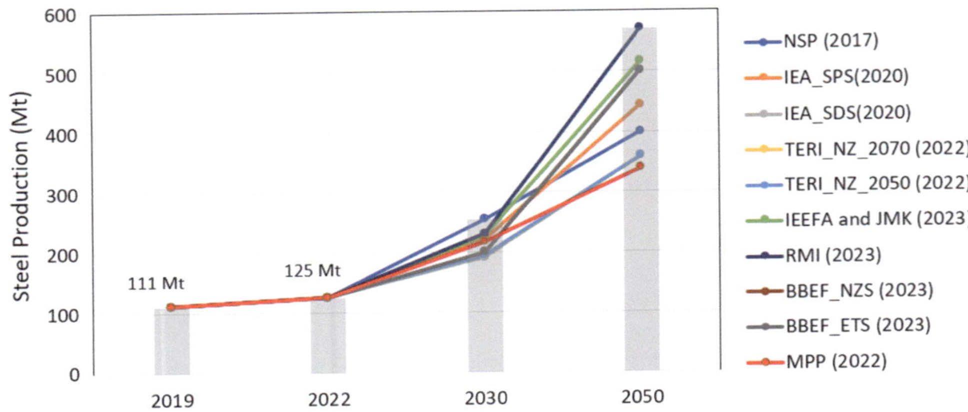

Fig. 7 Steel production in India is estimated to increase 2 × by 2030 and 4 × by 2050. [2, 24]. *Source* Author compilation based on NSP (2017), IEA, TERI (2022), [8], RMI, BBEF (2023), and MPP (2022). NSP—National Steel Policy, IEA SPS/SDS—International Energy Agency Stated Policy Scenario/Sustainable Development Scenario, TERI—The Energy and Resources Institute, IEEFA—Institute for Energy, Economics and Financial Analysis, JMK—JMK Research & Analytics, RMI—Rocky Mountain Institute, BBEF—Bloomberg New Energy Finance, MPP—Missions Possible Partnership

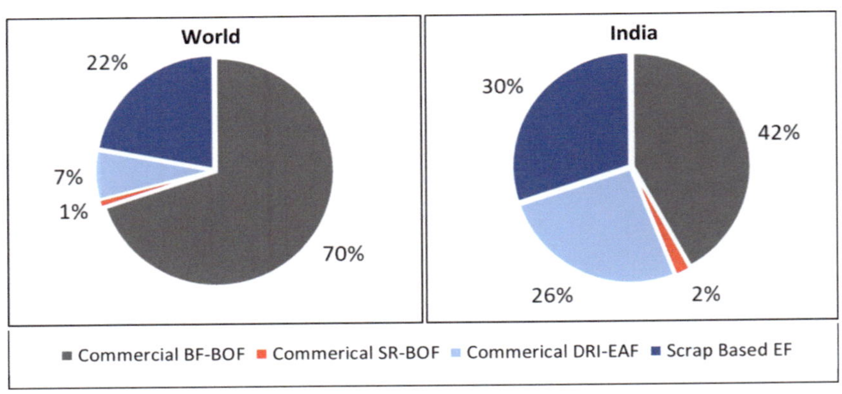

Fig. 8 Steelmaking by routes in the world (LHS) and India (RHS) in the year 2019 [2]. *Source* IEA Commercial BF-BOF includes traditional coal-, gas- and charcoal-based blast furnaces, with and without top-pressure recovery turbines, without CCUS. Commercial SR-BOF refers to smelting reduction without CCUS (COREX and FINEX). Commercial DRI-EAF includes gas- and coal-based DRI without CCUS, including that with a proportion of blended electrolytic hydrogen. Scrap-based EF refers to electric arc furnaces and induction furnaces fed mostly by scrap; it includes induction furnaces with a substantial iron charge

in the near future known as green premium. Higher prices can be compensated if the carbon prices are high enough to make conventional steel prices equal to the prices of green steel. Currently, however, the carbon prices are much lower than the expected green premium. This implies that the governments will have to support the green steel industry in the short to medium term.

There is a lot of uncertainty around the nature and structure of the industry. Whether the green steel will be produced by existing steel companies or companies from upstream value chains like mining or downstream value chains like capital goods companies that start producing their steel in captive plantsmuch like captive power plants. In addition, there are other challenges apart from the costs and investments needed for the transition to green steel. For instance, the size of the electrolyzer, the location of the electrolyzer, and the quality of iron ore. In India, low-grade variations make up a large majority of iron ore which may make it unsuitable for use in the green steel process, as these new technologies are understood to require an iron ore concentration of a minimum of 60%. The EAF route, which is important for transitioning toward green steel also uses high-grade iron ore [32].

However, despite these challenges and uncertainties, India can still capitalize on the green steel industry. This will need support from public policy in the form of incentives to steelmakers to produce green steel. India's steel sector is currently quite fragmented. The steel production capacity of about 143 Mt is spread across 878 plants with very few large producers and many small producers. Steel Authority of India Limited (SAIL), a public sector unit with 8 plants alone owns 14% of

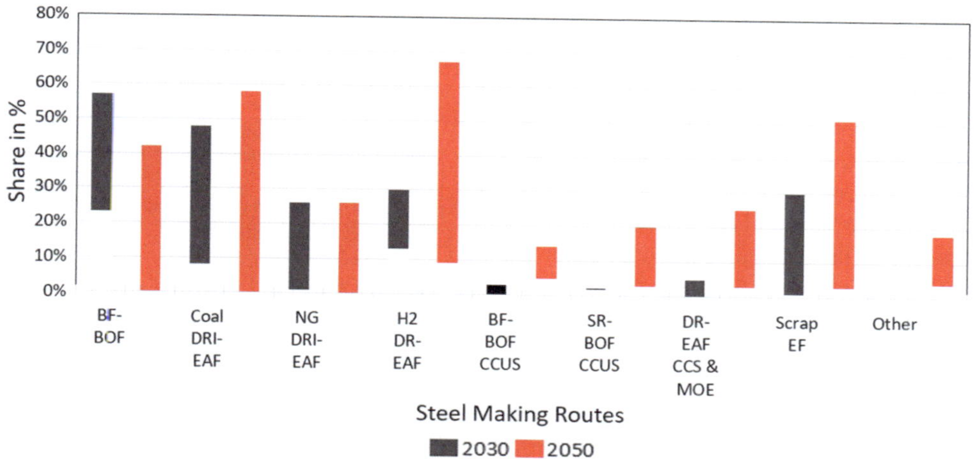

Fig. 9 Steelmaking routes and their share in total steel production in India by the years 2030 and 2050. *Source* Author compilation based on NSP (2017), IEA (2020), TERI (2022), RMI (2023), BBEF (2023), and MPP (2022). BF-BOF—Blast furnace-basic oxygen furnace; SR-BOF—Smelt reduction-basic oxygen furnace; Coal DRI-EAF—Coal direct reduced iron-electric arc furnace; NG DRI-EAF—Natural gas direct reduced iron-electric arc furnace; Scrap EF—Scrap electric furnace; H_2-DRI-EAF—Hydrogen direct reduced iron-electric arc furnace; BF-BOF CCUS—BF-BOF carbon capture utilization and storage; MOE—Molten oxide electrolysis

the capacity and produces 15% of crude steel [33]. Six large companies including SAIL, Rashtriya Ispat Nigam Limited (RINL), Tata Steel Group, ArcelorMittal Nippon Steel (AM/NS), Jindal Steel Power Limited (JSPL), and Jindal Steel Works Limited (JSW) produce about 60% of nation's crude steel. Twenty-six companies have 87% of the crude steel capacity of India and of this 11 of them have net-zero targets from the year 2040 to 2070. One way could be to incentivise the big companies to be lead the production of green steel. Once the production of green steel is scaled up, the costs will come down and then the small primary steelmakers can also adopt the technology, further driving the costs down. Large steel companies in India have already laid their capital expenditure plans with SAIL announcing capex of Rs 1 trillion over the next 10 years [34], JSW Steel of Rs 650 billion [35], and Tata Steel of Rs 160 billion [36], among others. Another alternative is to target the small and medium-sized steel companies with existing capabilities in steel production via coal-based direct reduced iron and natural gas-based direct reduced iron. These plants have expertise in DRI, an important aspect in H2-DRI using Indian iron ore grade and indigenous technologies, which can be used to kick-start hydrogen-based steelmaking [41]. Supporting the latter alternative might be more advantageous for India.

Currently, India has sectoral-level initiatives in the form of National Scrap Policy 2017, Perform, Achieve and Trade (PAT), New Energy and Industrial Technology Development Organization (NEDO) projects to increase the scrap availability at the domestic level, reduce energy consumption and enhance the energy efficiency in steelmaking. Through initiatives such as the Green Hydrogen Mission, the Indian steel sector has been allocated funds for research and development and deployment of hydrogen-based steelmaking in the sector [3]. However, this top-down approach may not yield large scale investments from the company. It requires a detailed bottom-up approach too, to understand the economic feasibility and the expected timelines so that appropriate policy interventions can be made. The European Union's introduction of a carbon border adjustment mechanism can be the signal to start.

In the absence of a strategic transition plan for the sector, new capacities will likely be created on the conventional technologies and locked-in fossil fuel energy sources. For India, this could result in a big miss on creating economic growth, clean jobs, energy security, and climate commitments.

5 Conclusions

Green steel is an opportunity for India. It has the potential to meet climate targets while cutting down imports of coking coal to reduce energy security concerns, create economic growth domestically and through, export markets and create clean jobs. Hydrogen-based steelmaking is emerging as one of the most promising technologies and there seems to be consensus that in India, H_2-DRI will be one of the major clean technologies producing steel, if not by 2030 then definitely by 2050. However, the policies and investment plans announced so far for the sector, cannot place India on the green steel pathway.

There are challenges in the sector as the technology is not commercial and cost uncertainty is high. However, the signals in the market for steel decarbonization are apparent starting with the carbon border tax. This can be a push for

India's steel sector to think of a low-emission strategy and commit investments but will need strong policy support. This is a new sector where no country has experience and so there will be learning-by-doing from the first few plants. India has the capability to undertake such efforts given its track record of deploying a solar capacity of 70 GW in less than a decade and creating a new solar industry by early engagement with private players. The key to unlocking the potential of green steel, therefore, is an early start. Collaboration with pilot projects in Europe could help in improving the current understanding of green steel. Global players like ArcelorMittal, Nippon Steel, POSCO, and SSAB, have been working on green steel technologies and projects that can impact the global steel industry. Parallelly investments are needed in the Indian plants to develop a comprehensive understanding on the feasibility and potential. Lessons from first few plants can help scale it in India and bring down the costs. Green steel, decarbonised steel, can be an economic growth story for India.

Acknowledgements We would like to thank Professor Ashish Garg, Professor Rajeev Jindal, Miss Anjali Gotekar and the participants of International Conference on Net Zero Emission Technologies for Sustainable Development: Challenges and Opportunities (N0ET—2022), IIT ISM Dhanbad, for their constructive comments and valuable feedback.

References

1. WSA (2022) World steel in figures 2022. World Steel in Figures 2022—worldsteel.org [3 February 2024]
2. IEA (n.d.) Global crude steel production by process route and scenario, 2019–2050. Global crude steel production by process route and scenario, 2019–2050—Charts—Data & Statistics—IEA [6 February 2024]
3. PIB Ministry of Steel (2023) 3 million tonne per annum capacity steel plant built at a cost of approximately ₹24000 crores. https://pib.gov.in/PressReleaseIframePage.aspx?PRID=1951911 [8 February 2024]
4. GoI (2017) National steel policy 2017. https://steel.gov.in/national-steel-policy-nsp-2017 [3 February 2024]
5. Sun W, Wang Q, Zhou Y, Wu J (2020) Material and energy flows of the iron and steel industry: Status quo, challenges and perspectives. Appl Energy 268:114946
6. WSA (2019) Global crude steel output increases by 3.4% in 2019. Global crude steel output increases by 3.4% in 2019—worldsteel.org [6 February 2024]
7. Ministry of Coal (2022) Export and import of coal. https://coal.gov.in/en/major-statistics/import-and-export [5 February 2024]
8. IEEFA, JMK (2023) Steel decarbonisation in India. Steel Decarbonisation in India_September 2023_2.pdf (ieefa.org) [10 March 2024]
9. Venkataraman M, Csereklyei Z, Aisbett E, Rahbari A, Jotzo F, Lord M, Pye J (2022) Zero-carbon steel production: the opportunities and role for Australia. Energy Policy 163:112811
10. Vogl V, Åhman M, Nilsson LJ (2018) Assessment of hydrogen direct reduction for fossil-free steelmaking. J Clean Prod 203:736–745
11. Wang R, Zhao Y, Babich A, Senk D, Fan X (2021) Hydrogen direct reduction (H-DR) in steel industry—an overview of challenges and opportunities. J Clean Prod 329:129797
12. Souza Filho IR, Ma Y, Kulse M, Ponge D, Gault B, Springer H, Raabe D (2021) Sustainable steel through hydrogen plasma reduction of iron ore: process, kinetics, microstructure, chemistry. Acta Mater 213:116971
13. Liberty Steel Group (n.d.) Pioneering GREENSTEEL through low carbon technologies. GREENSTEEL—LIBERTY Steel Group [15 February 2024]
14. Saarloha Kalyaani Group Company (n.d.) Kalyani Ferresta has a very low GHG emission. Kalyani Ferresta (saarloha.com) [1 March 2024]
15. AFRY (n.d.) Decarbonising the steel industry with new fossil-free production methods. Green Steel | AFRY [14 February 2024]
16. Leadership Group from Energy Transition (n.d.) Green steel tracker. Green Steel Tracker—Leadership Group for Industry Transition [6 February 2024]
17. H2 Green Steel (n.d.) Green steel production. Green Steel production— H2 Green Steel [15 February 2024]
18. SSAB (n.d.) Not all green steel is fossil-free steel—here's why. Green steel—Why is not all green steel fossil free?—SSAB [6 February 2024]
19. International Monetary Fund (2023) World economic outlook. World Economic Outlook (October 2023) (imf.org) [12 February 2024]
20. US Department of Commerce, Enforcement and Compliance (2023) Global steel trade monitor. Global Steel Trade Monitor [13 February 2024]
21. GoI, Department of Mines and Geology (n.d.) Metallic minerals. Metallic Minerals (rajasthan.gov.in) [12 February 2024]
22. WSA (2021) Apparent steel use per capita 2017 to 2021. World Steel in Figures 2022—worldsteel.org [14 February 2024]
23. JM BAXI Group (n.d.) Steel sector in India. Steel Sector In INDIA | J.M. Baxi Newsletter (jmbaxi.com) [9 February 2024]
24. Aayog N, RMI (2022) Harnessing green hydrogen opportunities for deep decarbonisation in India. Harnessing_Green_Hydrogen_V21_DIGITAL_29062022.pdf (niti.gov.in) [28 February 2024]
25. PV magazine (2023) JSW Steel to use green hydrogen to make steel in India. JSW Steel to use green hydrogen to make steel in India—pv magazine International (pv-magazine.com) [9 February 2024]
26. TATA Steel Europe Factbook (2020) HISARNA Building a sustainable steel industry. tata-steel-europe-factsheet-hisarna.pdf (tatasteeleurope.com) [1 March 2024]
27. Longden T, Beck FJ, Jotzo F, Andrews R, Prasad M (2022) 'Clean' hydrogen?—comparing the emissions and costs of fossil fuel versus renewable electricity based hydrogen. Appl Energy 306:118145
28. Howarth RW, Jacobson MZ (2021) How green is blue hydrogen? Energy Sci & Eng 9(10):1676–1687
29. Bhaskar A, Abhishek R, Assadi M, Somehesaraei HN (2022) Decarbonizing primary steel production: Techno-economic assessment of a hydrogen based green steel production plant in Norway. J Clean Prod 350:131339
30. Hall W, Kumar S, Kashyap S, Dayal S (2022) Achieving green steel: roadmap to a net zero steel sector in India. New Delhi: The Energy and Resources Institute (TERI)https://www.teriin.org/sites/default/files/files/Achieving_Green_Steel_Roadmap.pdf
31. Steel Orbis (2023) India and Sweden to collaborate on development and production of green steel. India and Sweden to collaborate on development and production of green steel (steelorbis.com) [5 February 2024]
32. RSC Group (2023) Steeldeal newsletter green steel- dreams vs reality: India's hurdles on the path to carbon- neutral steel. (23) Green Steel- Dreams vs Reality: India's Hurdles on the Path to Carbon- Neutral Steel | LinkedIn [1 March 2024]

33. PIB Ministry of Steel (2021) Total Steel Production Capacity in the country at present is 143.91 million tonnes. https://pib.gov.in/PressReleasePage.aspx?PRID=1736730 [12 February 2024]
34. The Hindu (2023) SAIL readying ₹1,00,000 cr capex plans. SAIL readying ₹1,00,000 cr capex plans—The Hindu BusinessLine [16 February 2024]
35. Market Today (2024) Rs 65,000 crore capex! JSW Steel shares jumped today. Rs 65,000 crore capex! JSW Steel shares jumped today. Key details—BusinessToday [16 February 2024]
36. Business Standard (2023) Tata Steel plans Rs 16,000 cr capex in FY24 for domestic, global operations. Tata Steel plans Rs 16,000 cr capex in FY24 for domestic, global operations (business-standard.com) [16 February 2024]
37. OCED International Energy Agency (2020) Iron and steel technology roadmap. https://www.iea.org/reports/iron-and-steel-technology-roadmap [3 February 2024]
38. Niti Aayog (2022) Harnessing green hydrogen, opportunities for deep decarbonisation in India. Harnessing_Green_Hydrogen_V21_DIGITAL_29062022.pdf (niti.gov.in) [2 February 2024]
39. International Trade Administration (2023) Global steel trade monitor. Global Steel Trade Monitor [7 February 2024]
40. Vogl V, Åhman M (2019) What is green steel?—towards a strategic decision tool for decarbonising EU steel', ESTAD proceedings: 4th ESTAD. Düsseldorf, Germany, 24–27 June
41. PIB Ministry of Steel (2023) Green steel. Press Information Bureau (pib.gov.in) [30 January 2024]

Energy Transition Modeling: Short-Term Electricity Demand Forecasting Using Seq2Seq Encoder–Decoder Model

Mantavya Upadhyay, Rajeev Jindal, and Vivek Tiwari

1 Introduction

Energy transition modeling is essential for effectively integrating renewable energy sources into the grid [1], and ensuring supply meets the demand. Accurately forecasting energy demand or consumption becomes necessary for planning, resource allocation, and seamless grid integration of renewable energy sources. Traditional forecasting methods include rule-based forecasting, wherein energy consumption is modeled as a function of factors like population [2]. The inability of rule-based forecasting to capture the complex relationships in the data [3] has led to the use of statistical methods like Auto-Regressive Integrated Moving Average (ARIMA) [4] and Seasonal Auto-Regressive Integrated Moving Averages (SARIMA) [5] for modeling energy consumption based on factors like the population and GDP of a country [6]. However, their need for stationarity and the restriction to modeling linear relationships [7] has led to the adoption of machine learning-based methods of forecasting based on linear regression [8, 9], artificial neural networks [10–12], and long short-term memory [13, 14] that has been widely used along with meteorological (temperature, humidity, rainfall, etc.) and economic indicators (GDP, per capita income, etc.). Despite their improved performance, the lack of hierarchical perception in shallow neural networks [15] limits their accuracy in tasks that require understanding complex patterns. Both ANNs and LSTMs consider the entire dataset at a time, which, in the context of time series forecasting, diminishes their ability to sequentially model the relationships between data points immediately preceding the forecast horizon [16]. XGBoost, an ensemble learning technique, excels in energy consumption forecasting due to its ability to handle complex, non-linear relationships and interactions among input features [17]. Transformer-based models, leveraging self-attention mechanisms, offer superior accuracy in capturing long-term dependencies and intricate patterns in time series data for energy demand forecasting [18].

Seq2Seq models, which were initially introduced for machine translation tasks [19], have been used for time series forecasting [13, 20] of univariate data because of their ability to capture the temporal dependencies of data. However, univariate models use only past values of the target variable. This limits the model's ability to capture influences from other relevant factors [21]. Data processing is vital in multivariate forecasting for improving data quality, feature extraction, normalization, and overall model accuracy [22, 23].

This paper investigates the effectiveness of various machine learning models for short-term electricity consumption forecasting. The study focuses on the Indian Institute of Technology Kanpur (IIT Kanpur) campus but also evaluates the models on publicly available datasets. The main objective is to identify the more accurate model for short-term electricity consumption forecasting by

- Comparing the performance of three machine learning models: Linear Regression, Artificial Neural Networks (ANNs), and Seq2Seq Encoder–Decoder models with LSTM unit cells.
- Highlighting the importance of weather features in forecasting electricity consumption accurately.

M. Upadhyay
Department of Electrical Engineering, Indian Institute of Technology, Kanpur, India

R. Jindal (✉) · V. Tiwari
Department of Sustainable Energy, Indian Institute of Technology, Kanpur, India
e-mail: rajeevj@iitk.ac.in

S. Narayanan et al. (eds.), *Materials, Devices and Systems for Sustainability*, IITK Directions 8,
https://doi.org/10.1007/978-981-96-7295-0_10

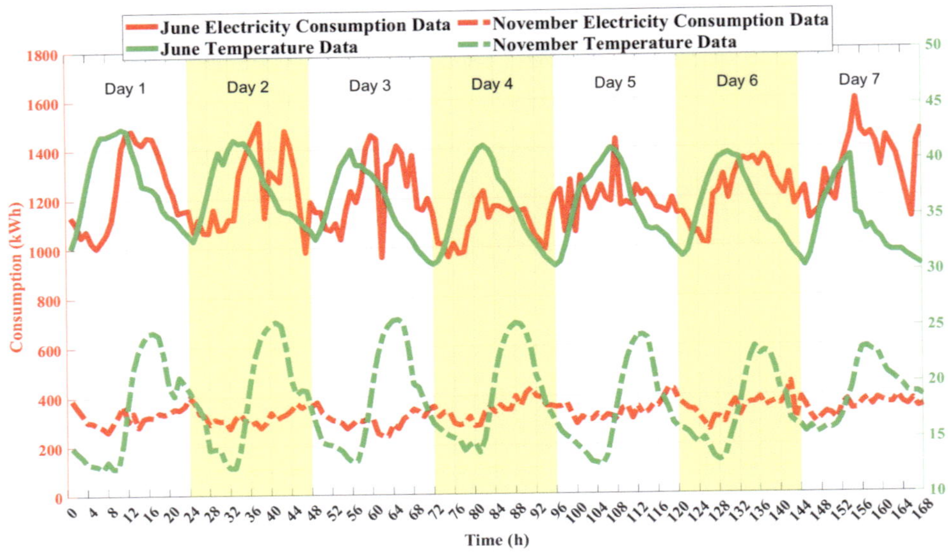

Fig. 1 Load versus temperature for a 7-day period (data is shown for November 11–17, 2022 and June 16–22, 2022) for the IIT Kanpur campus

2 Methodology

This section details the methodology used in this study to forecast electricity consumption. The methodology involved data collection, pre-processing, feature engineering, dimensionality reduction, and, finally, the application of machine learning models for forecasting.

2.1 Energy Consumption Dataset

In this study, the energy consumption dataset (with an interval of 5 min between November 1, 2022 and December 31, 2022) was obtained from the SG Centre, Indian Institute of Technology, Kanpur campus. For the comparison, the Spanish Electricity Network dataset [24] (Spanish electricity prices with an interval of 1 h, between January 1, 2014 to April 30, 2022) and the American Electric Power (AEP) dataset [25] (with an interval of 1 h, between December 31, 2004 to January 1, 2018) were also taken as a proof of concept for this study. The unit of the load consumption is kWh in the IIT Kanpur dataset, while that of power is MW for the AEP dataset.

2.2 Weather Data

The weather data for the IIT Kanpur geographical location (with an interval of 5 min between November 1, 2022 and December 31, 2022) was obtained from the Open-Meteo weather API [26], and contains the following parameters: wind speed (km/hr), temperature (° C), relative humidity (%), precipitation (mm), cloud cover (%), vapor pressure (kPa), soil temperature (° C), and direct normal irradiance (W/m^2). Similarly, the weather data for the geographical location of Madrid, the capital and the most populous city of Spain (with an interval of 1 h, January 1, 2014 to April 30, 2022) was obtained for the Spanish dataset.

Figure 1 plots the 7-day variation of electricity consumption and temperature (data is shown for November 11–17, 2022 and June 16–22, 2022) at the IIT Kanpur campus. It can be observed that the consumption values (in red) and the temperature values (in green) for June and November are seasonal. In June (summer) month, the highest and lowest recorded electricity consumption levels (for 5-min intervals) were 1594.03 kWh and 961.7 kWh, respectively. The temperature fluctuated between a peak of 43.1 °C and a low of 30.4 °C. In contrast, during November (winter), the maximum electricity demand drops by 70.5% (i.e., 470 kWh) and a minimum of 240.2 kWh. The temperature experienced its highest point at 25.8 °C and its lowest at 12.6 °C.

Furthermore, the peaks in temperature values are almost instantly followed by peaks in the consumption values. This can be inferred as a direct result of increased electricity consumption as the temperature increases in the day (10:00–14:00 h) due to increased cooling needs for the month of June. It validates the usage of weather features in forecasting electricity consumption.

2.3 Identifying and Removing Outliers

The energy consumption dataset (obtained from SG Centre, IIT, Kanpur) was visualized using Box Plot (Fig. 2) and Scatter Plot (Fig. 3) to analyze its distribution. The box plot (Fig. 2) demonstrates that the raw consumption data for the 5-min intervals has values greater than 10,000 kWh and 20,000 kWh for some data points.

Fig. 2 Box plot

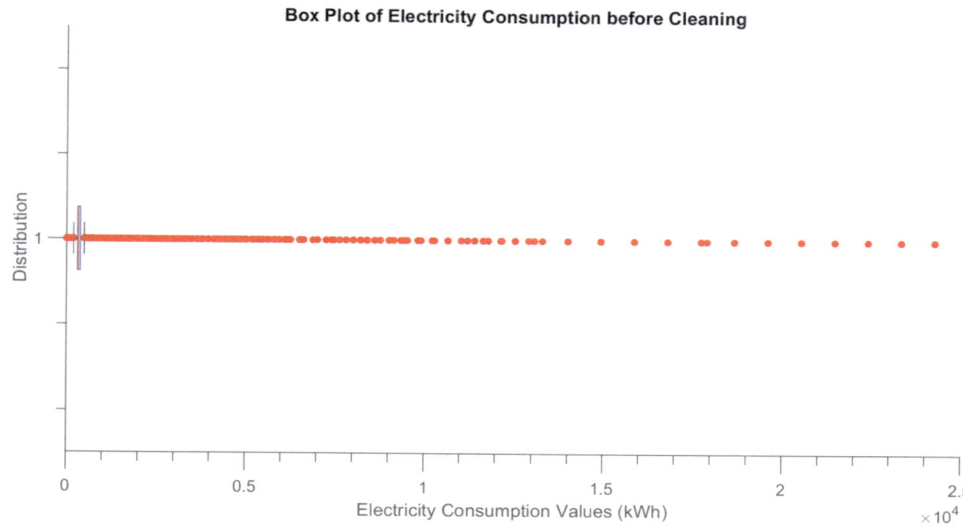

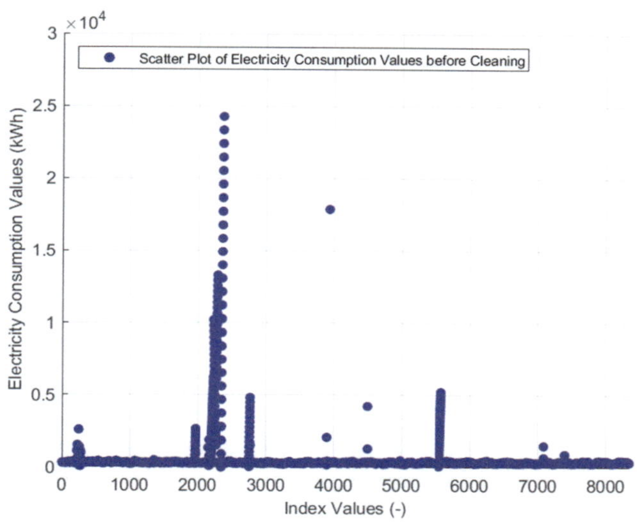

Fig. 3 Scatter plot

The scatter plot (Fig. 3) also demonstrates the same observation as the box plot; additionally, the scatter plot also gives us an idea of where these high values lie within the raw data. It was found that the raw data contains outliers [27] as the values are much larger than the mean and median of the dataset, and it was required to remove these outliers before feeding them into the machine learning models to increase their forecasting accuracy [27, 28]. Thus, to remove the outliers, the standard deviation method [27], inter-quartile range method [29], and the Z-score method [30] were used in this study.

Figure 4b displays the datasets after removing the outliers using various methods. It was noted that the IQR method was found to be most effective for removing outliers, resulting in a much more uniform distribution after the outliers are eliminated.

2.4 Feature Engineering

The dataset is now free from outliers. Several external features are now included to effectively represent the underlying dynamics of the time series data and capture the non-linear and temporal dependencies better [19]. These include lagged variables [31], weather variables [13], holidays [32] and seasonality indicators such as the day of the month, hours, and minutes [32]. The features that are included in this study are 34 weather features, 6 seasonality indicators, and 10 lagged features (50 in total).

2.5 Principle Component Analysis for Dimensionality Reduction

The dataset contains 50 features (discussed in the previous sub-section), including multiple lag and weather features such as direct incident irradiance, minimum and maximum daily temperatures, rainfall, and holidays. Training a model on such a dataset is computationally expensive. To address the challenge of high-dimensional feature spaces, Principal Component Analysis (PCA) [33] was employed in this study as PCA reduces the dimensionality of the dataset by transforming the original set of features into a smaller set of orthogonal variables.

Figure 5 shows the relationship between the cumulative explained variance (y-axis) and the number of features (x-axis). It was observed that the cumulative variance explained reaches saturation as the number of features increases. Beyond a certain number of features (18 in this case), the cumulative explained variance remains nearly the same, indicating that our model does not gain additional information from the rest of the features once saturation is

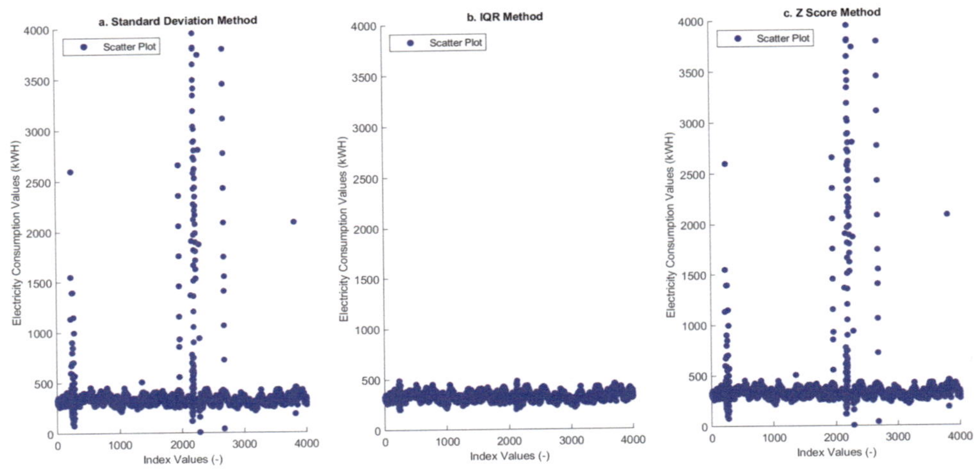

Fig. 4 Comparison of methods used for outlier removal

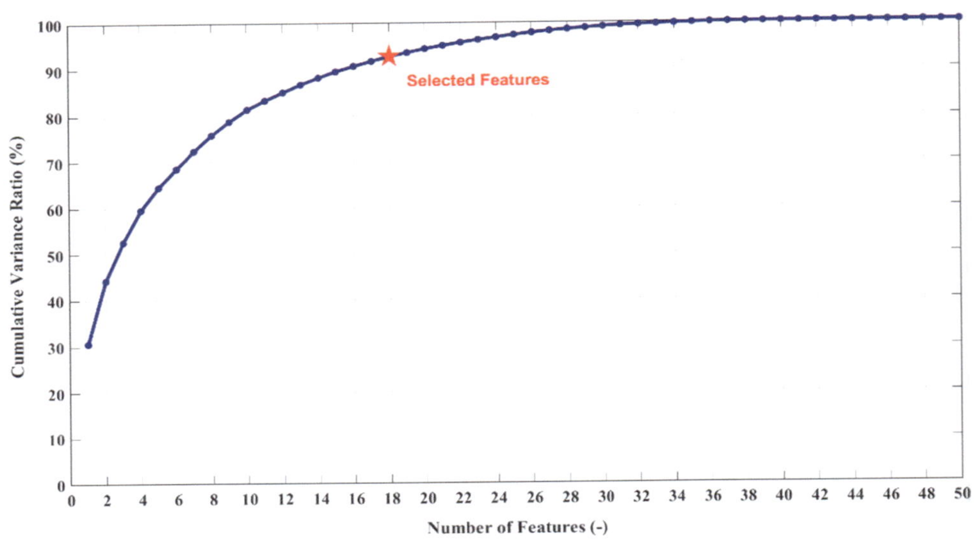

Fig. 5 Cumulative explained variance ratio versus the number of features

reached. There is a trade-off between slightly higher variance explained by using more than 18 features and the time required to train the model. Therefore, 18 features were used to train the forecasting model of this study.

2.6 Methods Used for Forecasting

In the present study, Multivariate Linear Regression (LR), Artificial Neural Networks (ANN), and Seq2Seq Encoder–Decoder Model with LSTM unit cell (Seq2Seq) methods are shortlisted (based on literature) and are discussed as follows:

- Linear regression is a statistical method for regression tasks [8]. It assumes that the relationship between the dependent (target) and independent (features) variables is linear and then fits a line or hyperplane according to the data [34]. It then makes predictions based on this best fit line. It is simple to understand and computationally less expensive, making it a widely used method. They are, however, not very reliable for big data and modeling non-linear data features in the energy datasets [35].
- Artificial Neural Networks (ANNs) are computational models that are inspired by the human brain and are used for various machine learning tasks, including regression and classification [24, 36]. ANNs consist of interconnected layers of nodes (neurons), where each connection has an associated weight. These networks learn by adjusting the weights based on the input data and prediction errors. ANNs are highly flexible and model complex, non-linear relationships between inputs and outputs. However, they are prone to overfitting and can only be used for generating point forecasts [37].
- Sequence-to-sequence (Seq2Seq) models with Long Short-Term Memory (LSTM) unit cells are a type of

neural network architecture that is used for tasks where the input and output are sequences, such as machine translation, text summarization, and time series forecasting [38, 39]. In this architecture, an encoder LSTM processes the input sequence and compresses that information into a fixed-length context vector. This context vector is then passed to a decoder LSTM, which generates the output sequence step by step. The use of LSTM units helps the model to remember long-term dependencies and handle vanishing gradient problems, making it particularly effective for sequences with temporal dependencies [14]. Seq2Seq models with LSTMs are well suited for time series forecasting because they can learn the patterns and trends in the input time series data and predict future values based on these learned patterns. Their ability to handle variable-length sequences and maintain memory over long sequences makes them powerful tools for capturing the complexities in time series data [20].

The above-shortlisted models are used to forecast the IIT Kanpur Campus's electricity consumption data and are also tested on two publicly available datasets in this study. Their results are discussed in the next section.

Some of the other forecasting models that have been compared against the proposed model in this study are Random Forest, XGBoost, GRU-based, SARIMA, and Transformers. They are briefly discussed here.

- Random Forest models are ensemble learning methods used for regression and classification tasks [40]. They operate by constructing multiple decision trees during training and outputting the mean prediction (regression) or mode of the classes (classification) of the individual trees [41]. Each tree in a random forest is built from a sample of the training data, and the features used for splitting nodes are selected randomly. This randomness helps to ensure that the individual trees are uncorrelated, leading to improved overall model performance and robustness [42].
- XGBoost, short for eXtreme Gradient Boosting, is a powerful and widely used machine learning algorithm for regression and classification tasks [43]. It belongs to the family of boosting algorithms, which iteratively combine weak learners (usually decision trees) to create a strong predictive model [17]. In the context of time series forecasting, XGBoost operates by iteratively refining its predictions based on the historical deviations between predicted and actual values [21]. It begins by fitting an initial model to the historical data and then sequentially adds decision trees, each trained on the residuals of the previous predictions. This approach allows XGBoost to capture complex temporal patterns and dependencies present in the time series data [17, 44].
- The Seasonal Autoregressive Integrated Moving Average (SARIMA) model is a statistical method widely used for time series forecasting, particularly when dealing with data that exhibits seasonal patterns and trends [5, 45]. SARIMA extends the traditional Autoregressive Integrated Moving Average (ARIMA) model by incorporating seasonal components [46].
- Transformers, originally designed for natural language processing, are increasingly applied to time series forecasting due to their ability to capture complex temporal patterns efficiently [18]. They use attention mechanisms for processing historical data, enabling them to capture both short- and long-term dependencies [25, 47]. They can also incorporate exogenous variables for enhanced forecasting [48].

2.7 Performance Metric and Horizons

In this study, some of the commonly used metrics [7, 49], including Mean Absolute Error (MAE), Mean Square Error (MSE), Root Mean Square Error (RMSE), and Symmetric Mean Absolute Percentage Error (sMAPE) metrics, were used to compare the accuracy of the forecasting models. Note that when comparing the performance of forecasting models, the lower values of MAE and other error metrics (MSE, RMSE, and sMAPE) are desirable [50, 51].

Table 1 provides description of the metrics used for evaluation in this study, where y_i are the actual values, $\widehat{y_i}$ are the predicted values, and n is the number of data points.

The model performances have been compared across 2 days and 1 week ahead of forecast horizons for the AEP dataset. The Spanish dataset has been evaluated on a forecast horizon of 1 h ahead. Forecasting across multiple horizons is crucial for selecting the model best suited for short- and long-term deployment. Three forecast horizons of 1, 12, and 72 time steps ahead have been evaluated for the IIT Kanpur dataset.

Table 1 Performance metric

Metric used	Definition
Mean absolute error (MAE)	$\text{MAE} = \frac{1}{n}\sum_{i=1}^{n}\left\lvert y_i - \widehat{y_i}\right\rvert$
Mean square error (MSE)	$\text{MSE} = \frac{1}{n}\sum_{i=1}^{n}\left(y_i - \widehat{y_i}\right)^2$
Root mean square error (RMSE)	$\text{RMSE} = \sqrt{\frac{1}{n}\sum_{i=1}^{n}\left(y_i - \widehat{y_i}\right)^2}$
Symmetric mean absolute percentage error (sMAPE)	$\text{sMAPE} = \frac{100\%}{n}\sum_{i=1}^{n}\frac{\left\lvert y_i - \widehat{y_i}\right\rvert}{\left(\lvert y_i\rvert + \left\lvert\widehat{y_i}\right\rvert\right)/2}$

3 Results

The selected models were trained and evaluated on 5-min interval electricity consumption data of IIT Kanpur with features representing the weather conditions (temperature, rainfall, direct solar irradiance), holidays, weekends, and lagged features to allow the models to learn from past consumption values.

3.1 Model Validation on Publicly Available Datasets

The performance of the proposed Seq2Seq encoder–decoder architecture has also been evaluated on two publicly available datasets, as mentioned in Sect. 2.1. Sections 3.1.1 and 3.1.2 compare the performance of popularly used forecasting models and the proposed Seq2Seq model.

3.1.1 Spanish Electricity Network Dataset

Table 2 represents the performance of linear regression, random forest [24], XGBoost [49], LSTM, and GRU-based models [16] in predicting the Spanish electricity prices as observed by [24], and the proposed Seq2Seq model. The performance of the proposed architecture has been evaluated with and without weather features to analyze the impact of weather features in forecasting energy demand.

It is observed that the proposed Seq2Seq model outperforms the most accurate model in [24] (LSTM) by over 78% in terms of MAE and over 45% in terms of RMSE. Furthermore, the model's performance improves by over 66% in MAE and 57% in RMSE when weather features are included in the forecasting compared to when they are not.

In Fig. 6, the performance of the Seq2Seq model is compared under two scenarios: (a) with weather features and (b) without weather features. It is observed that in Fig. 6a, the predicted values exhibit a dense clustering around the perfect prediction line, indicating lower errors in predictions. In Fig. 6b, it is observed that the distribution of the predictions deviates farther from the perfect prediction line, indicating higher prediction errors. This further validates the importance of using weather features for forecasting.

3.1.2 American Electric Power Dataset

The performance of the proposed architecture has also been evaluated using the publicly available American Electric Power (AEP) dataset. Table 3 represents the performance of SARIMA, Exponential Smoothing, Transformer, WaveNet, LSTM, tsRNN, Sparse Transformer, and Fixed Variant models on the AEP dataset, as observed by [25, 52], and the proposed Seq2Seq model for forecast horizons of 2 days and 7 days ahead, respectively. It should be noted that weather features have not been used for forecasting in the AEP dataset due to the challenge of selecting a single geographical location, given the vast expanse of the USA.

In Fig. 7, the performance of the Seq2Seq model is compared across two forecast horizons: (a) a 2-day-ahead forecast and (b) a 1-week-ahead forecast. In Fig. 7a, the predicted values are densely clustered around the perfect prediction line, indicating lower prediction errors. Conversely, in Fig. 7b, the distribution of predicted values is more scattered, indicating higher prediction errors. This difference in distribution is expected as shorter forecast horizons typically yield lower prediction errors than longer forecast horizons.

3.2 IIT Kanpur Data Set

Having established the Seq2Seq model's superior forecasting accuracy in Sect. 3.1, the study evaluates its performance on the IIT Kanpur dataset, which includes weather features, and compares its performance against commonly used models like linear regression and artificial neural networks to understand how the Seq2Seq model performs. The results of this comparison are discussed in this section.

3.2.1 Linear Regression Model

Table 4 presents the Ordinary Least Squares (OLS) regression analysis results for the dependent load variable (electricity

Table 2 Comparison of errors in forecasting the Spanish electricity prices

	Model	MAE	RMSE
Hernández Rodríguez et al. [24]	Linear regression	24.66	28.70
	Random forest	20.11	25.32
	XGBoost	10.69	14.40
	LSTM	9.17	12.79
	GRU	9.20	12.89
Current study	Linear regression	27.00	49.93
	Seq2Seq (without weather)	5.81	16.38
	Seq2Seq (with weather)	**1.95**	**7.01**

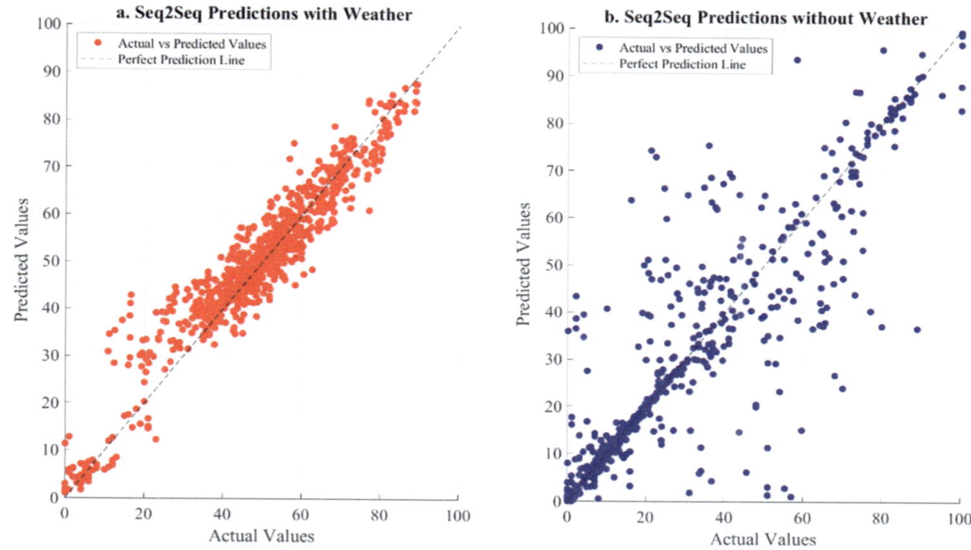

Fig. 6 Comparison of Seq2Seq model predictions on the Spanish electricity network dataset with and without weather features

Table 3 Comparison of errors in forecasting 2-day and 1-week-ahead values on the AEP dataset

	Model	AEP—2 days ahead			AEP—1 week ahead		
		RMSE (MW)	MAE (MW)	sMAPE (%)	RMSE (MW)	MAE (MW)	sMAPE (%)
Chan et al.	SARIMA	0.057	0.047	0.146	0.113	0.094	0.311
	Exponential smoothing	0.095	0.077	0.226	0.126	0.101	0.322
	Transformer	0.073	0.061	0.180	0.138	0.120	0.503
	WaveNet	0.053	0.044	0.134	0.088	0.073	0.257
	LSTM	0.114	0.100	0.303	0.121	0.102	0.352
	tsRNN	0.060	0.051	0.157	0.083	0.069	0.246
	Sparse transformer	0.060	0.051	0.179	0.084	0.070	0.236
	Fixed variant	0.060	0.051	0.177	0.084	0.070	0.235
Current study	Seq2Seq (without weather)	**0.027**	**0.013**	**0.053**	**0.034**	**0.019**	**0.073**

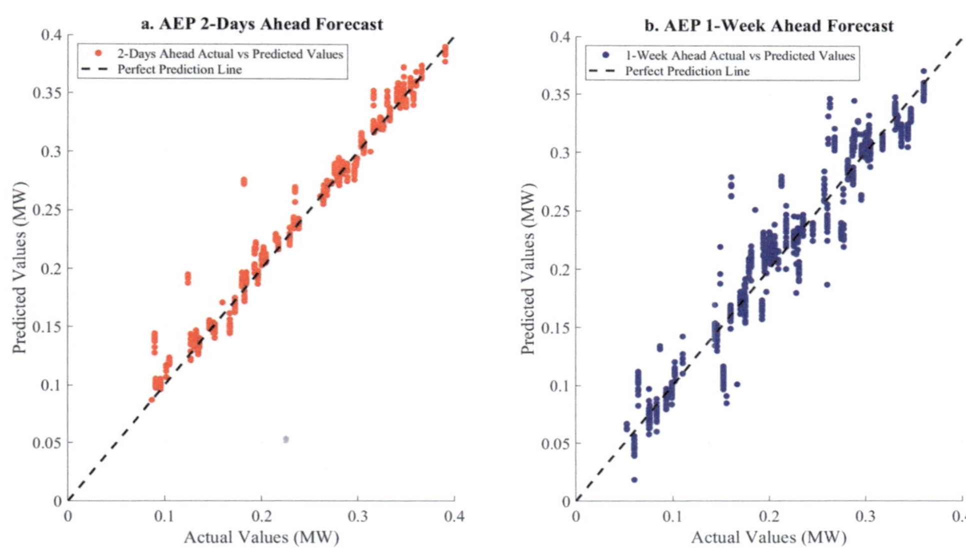

Fig. 7 Comparison of Seq2Seq model predictions on AEP dataset for **a** 2-day-ahead and **b** 1-week-ahead forecast horizons

Table 4 OLS regression results

Dep. variable	Load
R^2	0.250
Model	OLS
Adj. R^2	0.246
Method	Least Squares
F-statistic	64.29
Prob (F-statistic)	4.04×10^{-157}
Log-likelihood	−14,718
No. of observations	2719
AIC	2.947×10^4
Df residuals	2704
BIC	2.955×10^4
Df model	18
Covariance type	Non-robust

consumption). The regression analysis results indicate that the model explains 25% of the variance in the "Load" ($R^2 = 0.25$). The high F-statistic and the extremely low p-value indicate that the model is statistically significant. This means that at least one of the features included in the model has a significant linear relationship with the dependent variable "Load." Using non-robust covariance implies the assumption of constant variance (homoscedasticity) of error terms. However, the electrical consumption data does not have constant mean and variance, leading to high errors in forecasting by the linear regression model.

3.2.2 Artificial Neural Network

The Artificial Neural Network (ANN) model demonstrates an excellent ability to capture the general upward and downward trends in electricity consumption. This indicates that the ANN model is effective in learning the overall pattern and seasonal consumption behavior with respect to time. However, the model shows significant limitations in accurately predicting short-term fluctuations and struggles to forecast peak consumption values (refer to Fig. 8).

3.2.3 Seq2Seq Model

The Seq2Seq model excels in forecasting electricity consumption by effectively capturing both the overall trend and the short-term fluctuations in demand (refer to Fig. 9). Unlike the ANN model, the Seq2Seq model's architecture allows it to handle sequential data better, making it adept at identifying and predicting complex patterns in data [20]. This capability of accurately forecasting short-term fluctuations makes it apt for day-to-day operational planning and grid management.

Figure 10a is the scatter plot between the predicted load values from the linear regression model and the actual load values. The red line represents where the points would lie if the predictions were 100% accurate, i.e., the perfect prediction line. For data points above the red line, the predicted values exceed the actual values and vice versa. It can be observed that when the actual values are over 400, the linear regression model underestimates them, and when the actual values are less than 300, the model overestimates them. The predictions for the ANN model (Fig. 10b) tend to be slightly higher than the actual values, as indicated by the denser distribution above the perfect prediction line. However, in Fig. 10c, which depicts predictions from the Seq2Seq model, it can be observed that the bulk of the distribution is densely distributed around the perfect prediction line, indicating low error values in the predictions.

Table 5 shows the errors in forecasting by the three models for three different forecast horizons of 1, 12, and

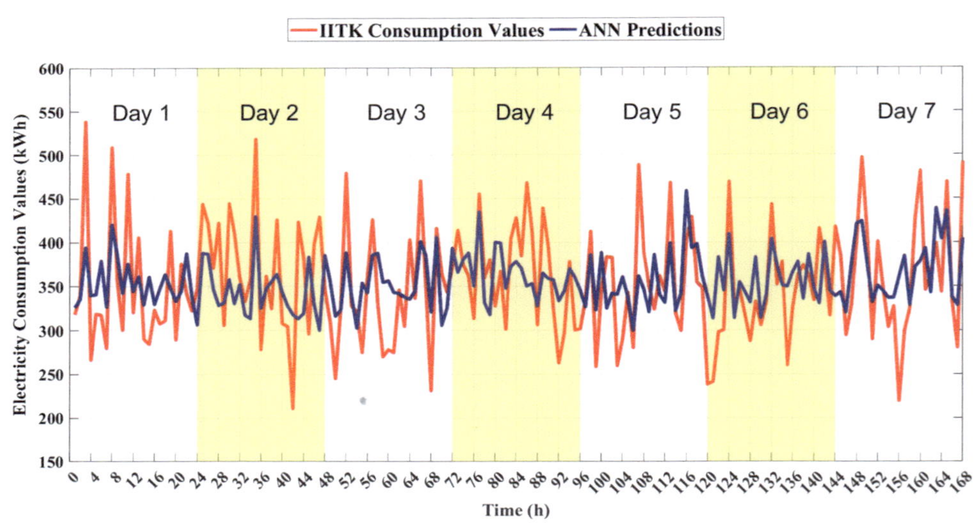

Fig. 8 Actual versus ANN model predictions on the IITK consumption data

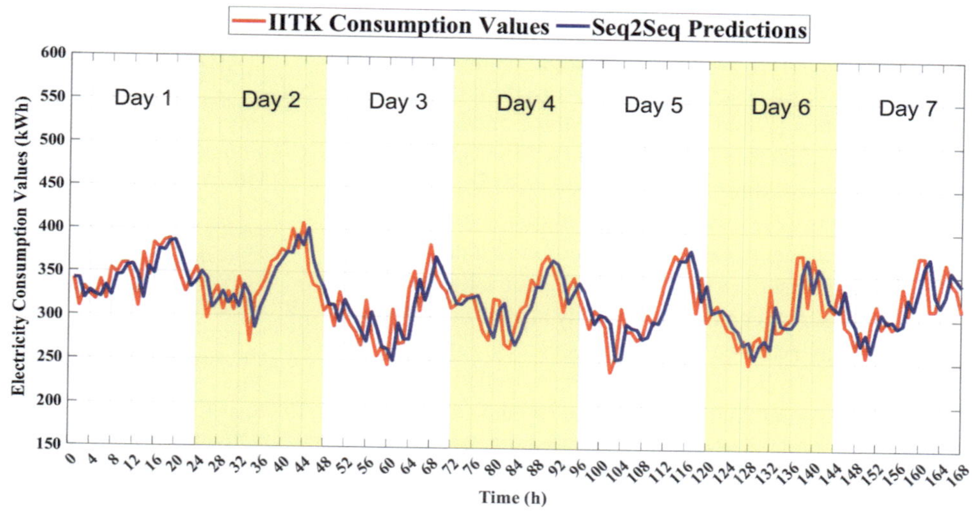

Fig. 9 Actual versus Seq2Seq model predictions on the IITK consumption data

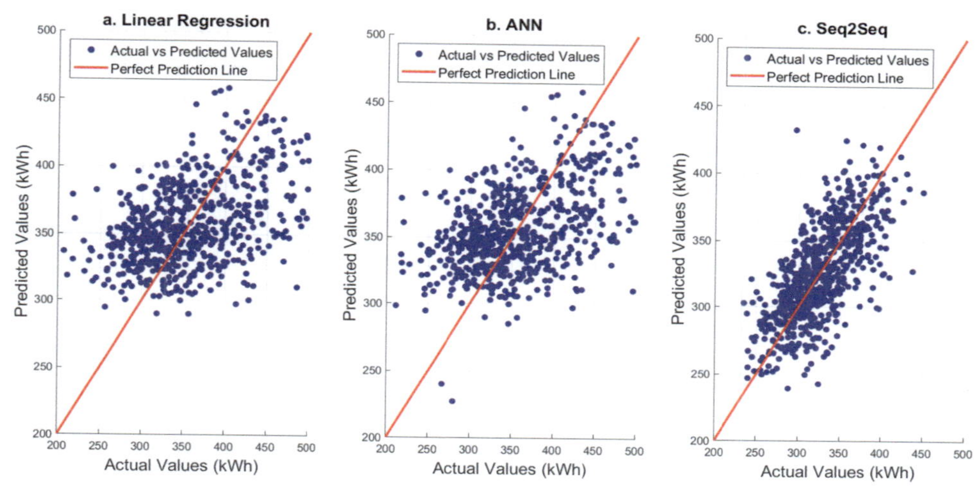

Fig. 10 Scatter plot between actual and predicted values using **a** Linear regression, **b** ANN, **c** Seq2Seq

Table 5 Comparison of errors for different forecast horizons

Models	Linear regression			Artificial neural network			Seq2Seq model		
Forecast horizon	MAE (kWh)	MSE (kWh^2)	RMSE (kWh)	MAE (kWh)	MSE (kWh^2)	RMSE (kWh)	MAE (kWh)	MSE (kWh^2)	RMSE (kWh)
1	50.49	5097.96	71.40	32.60	1767.61	42.043	1.70	58.86	7.67
12	93.93	17,687.64	132.99	46.56	3582.49	59.85	2.46	106.39	10.31
72	189.43	62,539.19	250.07	86.46	12,319.05	110.99	2.50	111.20	10.54

72 time steps, respectively, where each time step is 5 min long. It is observed that the proposed Seq2Seq model significantly outperforms both the linear regression and artificial neural networks in predicting the IIT Kanpur electricity consumption. The comparison of the performance of the linear regression, artificial neural network, and Seq2Seq model on the IIT Kanpur dataset demonstrates the superior accuracy of prediction of the Seq2Seq model. The Seq2Seq model outperforms both models by over 90% (MAE) across all three forecast horizons. Thus, it can be inferred that the proposed Seq2Seq model is suitable for electricity consumption forecasting [50].

4 Conclusion

This study proposed the data processing framework (involving outlier removal, feature engineering, and dimensionality reduction) and compared the effectiveness of different machine learning models for short-term electricity forecasting. It analyzed three datasets: one from the Indian Institute of Technology Kanpur (IIT Kanpur) campus, another from the Spanish Electricity Network, and a third from American Electric Power and focused on three models: Linear Regression, Artificial Neural Networks (ANNs), and Seq2Seq models with Long Short-Term Memory (LSTM) units. The first two datasets were used for model validation to choose the most accurate model (found to be the seq2seq model), and the third dataset was used for forecasting the electricity consumption at the IIT Kanpur campus. The study also investigated the impact of weather data on forecasting accuracy, validating the use of weather features for more accurate predictions.

The study found that the Seq2Seq model significantly outperformed both models by more than 90% (MAE) in predicting electricity consumption. This is because the Seq2Seq model can capture both long-term trends and short-term fluctuations in energy demand. This makes it a promising tool for improving electricity demand forecasting. Future research using the seq2seq model can explore incorporating additional feature sources, such as demographic details (changes in faculty, staff, and student populations) and growth in electricity consumption per capita along with the changes in per capita GDP, for long-term forecasting of the electricity consumption until the year 2030.

Acknowledgements The authors would like to thank the SG Centre, IIT, Kanpur for sharing the Energy Consumption Dataset. The authors would also like to thank Prof. Swathi Battula of the Electrical Engineering Department at IIT Kanpur for her valuable input throughout the project.

Dataset Availability The Spanish Electricity Network dataset can be found at https://github.com/eleion/Papers/tree/main/Energy%20pricing and the AEP dataset at https://www.kaggle.com/datasets/robikscube/hourly-energy-consumption/data.

References

1. Phoumin H, Kimura F, Arima J (2021) Asean's energy transition towards cleaner energy system: energy modelling scenarios and policy implications. Sustain 13(5):1–29. https://doi.org/10.3390/su13052819
2. Ilin IV, Iliashenko OY, Schenikov EM (2021) An approach to forecasting the structure of energy generation in the age of energy transition based on the automated determination of factor significance. Energies 17(1). https://doi.org/10.3390/en17010068
3. Taylor JW, de Menezes LM, McSharry PE (2006) A comparison of univariate methods for forecasting electricity demand up to a day ahead. Int J Forecast 22(1):1–16. https://doi.org/10.1016/j.ijforecast.2005.06.006
4. Chodakowska E, Nazarko J, Nazarko Ł (2023) Arima models in electrical load forecasting and their robustness to noise. Energies 14(23). https://doi.org/10.3390/en14237952
5. Deretić N, Stanimirović D, Al Awadh M, Vujanović N, Djukić A (2022) SARIMA modelling approach for forecasting of traffic accidents. Sustain 14(8). https://doi.org/10.3390/su14084403
6. Mauleón I (2022) A statistical model to forecast and simulate energy demand in the long-run. Smart Energy 7(July). https://doi.org/10.1016/j.segy.2022.100084
7. Kontopoulou VI, Panagopoulos AD, Kakkos I, Matsopoulos GK (2023) A review of ARIMA versus machine learning approaches for time series forecasting in data driven networks. Futur. Internet 15(8):1–31. https://doi.org/10.3390/fi15080255
8. Ciulla G, D'Amico A (2019) Building energy performance forecasting: A multiple linear regression approach. Appl Energy 253(June):113500. https://doi.org/10.1016/j.apenergy.2019.113500
9. Braun MR, Altan H, Beck SBM (2014) Using regression analysis to predict the future energy consumption of a supermarket in the UK. Appl Energy 130:305–313. https://doi.org/10.1016/j.apenergy.2014.05.062
10. Kaur K (2023) Artificial neural network model to forecast energy consumption in wheat production in India. J Stat Theory Appl 22(1–2):19–37. https://doi.org/10.1007/s44199-023-00052-w
11. Runge J, Zmeureanu R (2019) Forecasting energy use in buildings using artificial neural networks: a review. Energies 12(17). https://doi.org/10.3390/en12173254
12. Taleb I, Guerard G, Fauberteau F, Nguyen N (2022) A flexible deep learning method for energy forecasting. Energies 15(11). https://doi.org/10.3390/en15113926
13. Stefenon SF, Seman LO, Aquino LS, Coelho L dos S (2023) Wavelet-Seq2Seq-LSTM with attention for time series forecasting of level of dams in hydroelectric power plants. Energy 274(March):127350. https://doi.org/10.1016/j.energy.2023.127350
14. Bareth R, Yadav A, Gupta S, Pazoki M (2024) Daily average load demand forecasting using LSTM model based on historical load trends. IET Gener Transm Distrib 18(5):952–962. https://doi.org/10.1049/gtd2.13132
15. Marcus G (2018) Deep learning: a critical appraisal, pp 1–27 [Online]. http://arxiv.org/abs/1801.00631
16. Al-Selwi SM, Hassan MF, Abdulkadir SJ, Muneer A (2023) LSTM inefficiency in long-term dependencies regression problems. J Adv Res Appl Sci Eng Technol 30(3):16–31. https://doi.org/10.37934/araset.30.3.1631
17. M'hamdi O, Takács S, Palotás G, Ilahy R, Helyes L, Pék Z (2024) A comparative analysis of XGBoost and neural network models for predicting some tomato fruit quality traits from environmental and meteorological data. Plants 13(5). https://doi.org/10.3390/plants13050746
18. Oliveira HS, Oliveira HP (2023) Transformers for energy forecast. Sensors 23(15):1–21. https://doi.org/10.3390/s23156840
19. Sutskever I, Vinyals O, Le QV (2014) Sequence to sequence learning with neural networks. Adv Neural Inf Process Syst 4(January):3104–3112
20. Masood Z, Gantassi R, Ardiansyah, Choi Y (2022) A multi-step time-series clustering-based Seq2Seq LSTM learning for a single household electricity load forecasting. Energies 15(7). https://doi.org/10.3390/en15072623
21. Hadri S, Najib M, Bakhouya M, Fakhri Y, El Arroussi M (2021) Performance evaluation of forecasting strategies for electricity consumption in buildings. Energies 14(18). https://doi.org/10.3390/en14185831
22. Bhatnagar R (2018) Machine learning and big data processing: a technological perspective and review. Adv Intell Syst Comput 723(January):468–478. https://doi.org/10.1007/978-3-319-74690-6_46
23. Kumar Padhi B, Nayak SS, Biswal BN (2018) Machine learning for big data processing: a literature review. Int J Innov Res Technol IJIRT 140001 December 2018|IJIRT 5(7):10 | ISSN 2349–6002 samples [Online]. https://www.researchgate.net/publication/331497335

24. Hernández Rodríguez M, Baca Ruiz LG, Criado Ramón D, Pegalajar Jiménez M del C (2023) Artificial intelligence-based prediction of Spanish energy pricing and its impact on electric consumption. Mach Learn Knowl Extr 5(2):431–447. https://doi.org/10.3390/make5020026
25. Chan JW, Yeo CK (2024) A transformer based approach to electricity load forecasting. Electr J 37(2):107370. https://doi.org/10.1016/j.tej.2024.107370
26. Open Meteo Weather API. Accessed: May 01, 2024. [Online]. https://open-meteo.com/
27. Cousineau D, Chartier S (2010) Outliers detection and treatment: a review. Int. J. Psychol. Res. 3(1):58–67. https://doi.org/10.21500/20112084.844
28. Lee H, Yun S (2024) Strategies for imputing missing values and removing outliers in the dataset for machine learning-based construction cost prediction. Buildings 14(4):933. https://doi.org/10.3390/buildings14040933
29. Carifio J (2006) Standard scores based on the median and inter-quartile range. J Math Stat 2(1):367
30. Dastjerdy B, Saeidi A, Heidarzadeh S (2023) Review of applicable outlier detection methods to treat geomechanical data. Geotechnics 3(2):375–396. https://doi.org/10.3390/geotechnics3020022
31. Surakhi O et al (2021) Time-lag selection for time-series forecasting using neural network and heuristic algorithm. Electron 10(20):1–22. https://doi.org/10.3390/electronics10202518
32. López M, Sans C, Valero S, Senabre C (2019) Classification of special days in short-term load forecasting: the Spanish case study 12(7). https://doi.org/10.3390/en12071253
33. Kitao A (2022) Principal component analysis and related methods for investigating the dynamics of biological macromolecules. J 5(2):298–317. https://doi.org/10.3390/j5020021
34. Smedegård OØ, Jonsson T, Aas B, Stene J, Georges L, Carlucci S (2021) The implementation of multiple linear regression for swimming pool facilities: case study at Jøa, Norway. Energies 14(16):1–23. https://doi.org/10.3390/en14164825
35. Deb C, Zhang F, Yang J, Lee SE, Shah KW (2017) A review on time series forecasting techniques for building energy consumption. Renew Sustain Energy Rev 74(February):902–924. https://doi.org/10.1016/j.rser.2017.02.085
36. Pavićević M, Popović T (2022) Forecasting day-ahead electricity metrics with artificial neural networks. Sensors 22(3). https://doi.org/10.3390/s22031051
37. Wang H, Lei Z, Zhang X, Zhou B, Peng J (2019) A review of deep learning for renewable energy forecasting. Energy Convers Manag 198(July):111799. https://doi.org/10.1016/j.enconman.2019.111799
38. Gong G, An X, Mahato NK, Sun S, Chen S, Wen Y (2019) Seq2seq model. Energies 12(16):1–18. https://doi.org/10.3390/en12163199
39. Li Y, Du M, He S (2022) Attention-based sequence-to-sequence model for time series imputation. Entropy 24(12). https://doi.org/10.3390/e24121798
40. Dudek G (2022) A comprehensive study of random forest for short-term load forecasting. Energies 15(20). https://doi.org/10.3390/en15207547
41. Bento P, Pombo J (2024) Forest and optimal feature selection. Energies. https://doi.org/10.3390/en17081926
42. Tyralis H, Papacharalampous G (2017) Variable selection in time series forecasting using random forests. Algorithms 10(4). https://doi.org/10.3390/a10040114
43. Ahn JM, Kim J, Kim K (2023) Ensemble machine learning of gradient boosting (XGBoost, LightGBM, CatBoost) and attention-based CNN-LSTM for harmful algal blooms forecasting. Toxins (Basel), 15(10). https://doi.org/10.3390/toxins15100608
44. Saigustia C, Pijarski P (2023) Time series analysis and forecasting of solar generation in Spain using eXtreme gradient boosting: a machine learning approach. Energies 16(22). https://doi.org/10.3390/en16227618
45. Costa GE de M e, Menezes Filho FCM d, Canales FA, Fava MC, Brandão ARA, Paes RP de (2023) Assessment of time series models for mean discharge modeling and forecasting in a sub-basin of the Paranaíba river, Brazil. Hydrology 10(11):1–20. https://doi.org/10.3390/hydrology10110208
46. Tang C, Tong Z, Wei Y, Wu X, Tian X, Yang J (2023) Time–frequency characteristics and SARIMA forecasting of atmospheric water Vapor in East Asia. Atmosphere (Basel) 14(5):1–24. https://doi.org/10.3390/atmos14050899
47. Zhao Z et al (2021) Short-term load forecasting based on the transformer model. Inf 12(12):1–22. https://doi.org/10.3390/INFO12120516
48. Liu X, Wang W (2024) Deep time series forecasting models: a comprehensive survey. Mathematics 12(10):1504. https://doi.org/10.3390/math12101504
49. Ji Q, et al (2022) Short-and medium-term power demand forecasting with multiple factors based on multi-model fusion. Mathematics 10(12). https://doi.org/10.3390/math10122148
50. Cummings DAT (2017) Relative mean absolute error 70(3):285–292. https://doi.org/10.1080/00031305.2016.1148631.Case
51. Maçaira PM, Cyrino Oliveira FL (2016) Another look at SSA. Boot forecast accuracy. Int J Energy Stat 04(02):1650008. https://doi.org/10.1142/s2335680416500083
52. Liu Y, Dutta S, Kong AWK, Yeo CK (2023) An image inpainting approach to short-term load forecasting. IEEE Trans Power Syst 38(1):177–187. https://doi.org/10.1109/TPWRS.2022.3159493

Policy and Regulatory Drivers for Renewable Energy Development in India

Anoop Singh, Madhav Sharma, Sonam Dolker, and Shambhavi Mishra

1 Introduction

The Industrial Revolution and subsequent phases of industrial development in the nineteenth and early twentieth centuries witnessed a global energy surge, fuelled primarily by coal and petroleum, the dominant sources of the time [1, 2]. This fossil-fuel-based surges in energy consumption came at a high environmental cost. Greenhouse gas (GHG) emissions, particularly carbon dioxide, exhibited a compound annual growth rate (CAGR) of 1.756% from 1990 to 2022 [3] (Fig. 1), sparking concerns about global warming and climate change [4]. Faced with this stark reality, nations worldwide were compelled to reconsider their energy strategies.

The power sector, a significant contributor since the Industrial Revolution, faced an escalating carbon footprint due to a growing population and rising per capita energy consumption [5]. Decarbonizing this sector became paramount, propelling cleaner alternatives like solar and wind power to the forefront. However, these nascent RE sources faced significant hurdles. Initially, they lacked cost-effectiveness and economies of scale compared to established fossil fuel technologies. Policy and regulatory interventions, therefore, became crucial to level the playing field, making renewables more competitive [6].

India's embrace of renewables stemmed initially from the widening energy access gap [7]. Millions in remote locations lack grid connectivity, and even semi-urban areas struggle with frequent power shortages [8]. To bridge this divide and address these disparities, India needed a solution that could reach the socio-economically deprived across rural and remote areas of the country [9]. Renewable Energy Sources (RES) offered a solution to electrify such a location, albeit with limited hours of supply and operational challenges, alongside improved reliability [10, 11].

Various policy and regulatory developments have played a pivotal role in propelling the promotion of RE [6, 12]. Building upon the foundation laid by the Electricity Act 2003 [13, 14], subsequent policies and regulations such as the National Electricity Policy (NEP), Tariff Policy (TP), Policy for Standalone Systems, and Rural Electrification Policy (REP) have significantly reinforced India's dedication to ensuring energy access and fostering the adoption of RE [8].

Strategic initiatives like Renewable Purchase Obligations (RPOs), Renewable Energy Certificates (RECs) [15, 16], GTAM and GDAM, and Renewable Generation Obligations (RGOs), along with state-level grid-interactive rooftop solar and minigrid policies, have further cemented India's position as a global leader in RE deployment [17]. These measures not only underscore India's commitment to sustainable energy but also establish ambitious targets for the future, including its Intended Nationally Determined Contribution (INDC) commitments [18]

India's stringent policy measures, illustrated by the initiatives discussed above, have been instrumental in the adoption of RES [17]. The share of RES was 10.55% (16.8 GW of 159.3 GW) in 2009–2010 [19], and it has grown to 31.26% in 2023–2024 (152.1 GW of 429.9 GW) [20, 21]. Driven by falling cost and competitive bidding, the increasing penetration of

A. Singh (✉) · M. Sharma · S. Mishra
Centre for Energy Regulation (CER) and Energy Analytics Lab (EAL), Department of Management Sciences (DoMS), Indian Institute of Technology Kanpur, Kanpur, India
e-mail: anoops@iitk.ac.in

M. Sharma
e-mail: madhavsh@iitk.ac.in

S. Mishra
e-mail: sbmishra@iitk.ac.in

A. Singh · S. Dolker
Department of Sustainable Energy Engineering (SEE), Indian Institute of Technology Kanpur, Kanpur, India
e-mail: sonamd21@iitk.ac.in

S. Narayanan et al. (eds.), *Materials, Devices and Systems for Sustainability*, IITK Directions 8,
https://doi.org/10.1007/978-981-96-7295-0_11

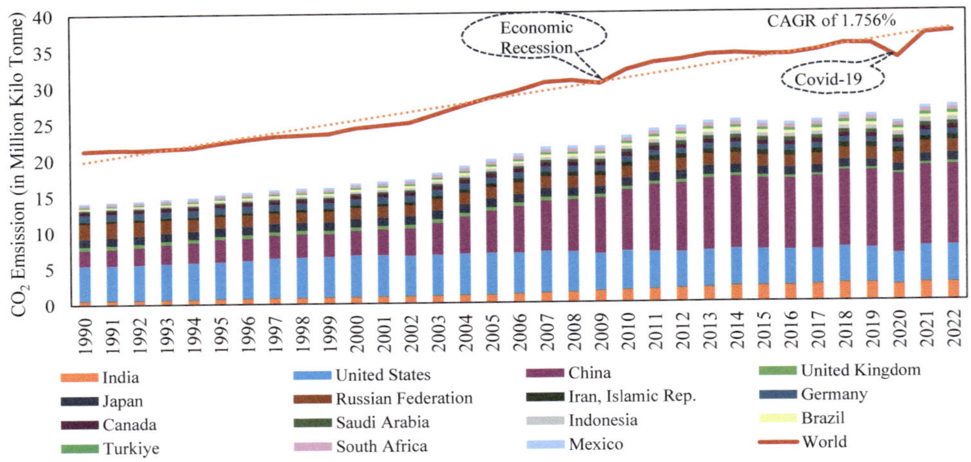

Fig. 1 Global trends of CO_2 emissions in last three decades [3]

solar energy has fueled RE growth. However, RE generation is seasonal and exhibits variability and uncertainty posing challenges for planning for resource adequacy and power system operation [22].

This paper presents an evolutionary journey of India's evolving policy and regulatory landscape for RE, exploring its transition from prioritizing energy access toward achieving long-term sustainability and reliability. The paper then analyzes the impact of key policies, regulations, plans, and schemes on RE adoption. It delves into the challenges hindering broader RE adoption, such as variability and intermittency, and discusses the role of better forecasting and energy storage integration. Subsequent sections provide a review of the impact of implementing RE policies and regulations in India.

2 Evolution of RE Growth in India

In the early stages of India's development, RE sources, then known as non-conventional sources, attracted the attention of the planners and policymakers. The Department of Science and Technology (DST) established the Commission for Additional Sources of Energy (CASE) in 1981 to focus on these resources. The commission swiftly evolved as a dedicated Department of Non-Conventional Energy Sources (DNES) in 1982. By 1992, it became the Ministry of Non-Conventional Energy Sources (MNES), marking India as the first country with a dedicated ministry for RE development.

The policies of the 1980s and 1990s focused more on supporting small-scale RE projects through various government programs. A key policy introduced in the early 1990s, pushed by private investors, was similar to today's Feed-in Tariff (FiT), which facilitated the procurement of excess RE generation. This policy significantly contributed to the growth of grid-connected wind generators, especially in the southern region of the country. Additionally, the Electricity Act of 2003 was a landmark development that provided a comprehensive framework for promoting RE.

The Generation-Based Incentives (GBI) scheme launched in 2011 further boosted RE adoption by incentivizing generation from wind and solar projects [23]. Additionally, accelerated depreciation offered tax benefits, making RE projects attractive from taxation point of view. Recognizing the crucial role of grid integration, CERC regulations granted "must-run" status to RE projects (excluding biomass), minimizing curtailment and providing stability for RE generation [24, 25].

The establishment of the National Action Plan on Climate Change (NAPCC) in June 2008 marked a crucial stepping stone toward mitigating and adapting to the extreme effects of climate change [26]. At United Nations Climate Change Conference (COP21) in 2015, India committed to achieve 40% of installed capacity from non-fossil fuel sources by 2030 and reduce emission intensity of its GDP by 33–35% from 2005 level as part of its Nationally Determined Contributions (NDCs) [27]. In 2021 at COP26, India introduced its "Panchamrit" commitments, aiming for significant reductions in emissions, increased RE share, net-zero emissions by 2070, and 500 GW of non-conventional energy by 2030 [18].

Driven by economies of scale and technological advancements, levelized costs of electricity for solar PV, onshore wind, and offshore wind technologies have seen reductions of 89%, 69%, and 59%, respectively, over 2010–2022. This reduction mirrors a global trend of significant declining cost of solar and wind technologies [28]. The enacted policies and further cost reduction of RE technology have yielded larger adoption. Since 2007–2008, India has witnessed a significant growth in RE deployment (Fig. 2). While thermal power growth has slowed, RE additions have steadily increased, showcasing a significant shift in the energy landscape. This translates to a substantial rise in RE capacity share, reaching 31.26% (42.21% including large hydro) as of December 31, 2023.

Fig. 2 Year-wise growth of installed capacity and RE share [20, 21]

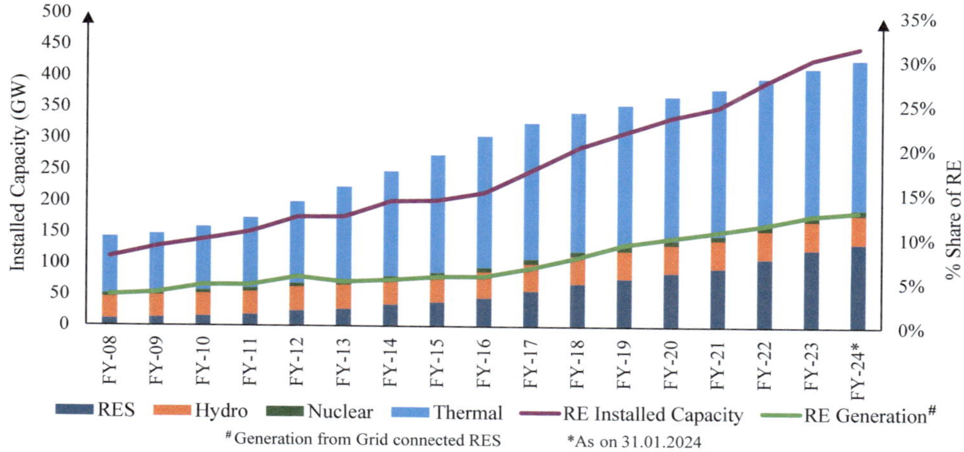

2.1 Evolution of Policy and Regulatory Framework

India's recent surge in RE adoption is rooted in the reforms initiated during the 1990s, coinciding with the country's economic liberalization. The power sector underwent significant changes, with unbundling State Electricity Boards (SEBs) and allowing independent entities to generate, transmit, and distribute electricity [8, 16]. Private investments, attracted by government incentives like accelerated depreciation, tax holidays, and relaxed foreign investment procedures, played a pivotal role in development of the power sector. States such as Andhra Pradesh, Karnataka, and Tamil Nadu pioneered guaranteed minimum purchase rates, providing developers with assured returns [29].

Furthermore, Electricity Act (EA) 2003 provided a provision and extensive framework to promote RE adoption. Section 86 of the Act mandated RPOs for the distribution companies, ensuring a fixed minimum purchase of RE (Table 1). It introduced guidelines for RE procurement through competitive bidding as per Section 63 [8]. The obligation applies to open-access consumers as well as consumers with captive generation capacity. The Act empowers rural electrification by exempting licensing requirements for local authorities and promoting non-discriminatory access to the grid (Sect. 5/14). This legislation laid the foundation for India's RE growth, making it a cornerstone of the country's clean energy transition.

Subsequent policies, including the National Electricity Policy (NEP) 2005 and the Tariff Policy 2006 and 2016, reinforced India's commitment to a sustainable RE ecosystem. These policies promoted private sector participation, introduced competitive bidding, and enabled cross-subsidies for underprivileged consumers. However, an emphasis on collaboration among all stakeholders is required for sustainable and inclusive growth.

Beyond top-down policies, it advocates a role of minigrids through a delicensed ecosystem. Incentivizing both production and consumption is deemed essential for India's continuous progress toward a sustainable and equitable transition to an RE future. Green electricity products are now part of the electricity markets enabling a number of distribution utilities to participate which are otherwise restricted by their geographies.

2.2 RE Programmes and Achievements Thereof

India's RE landscape has undergone a transformation, driven by the strategic implementation of policy and regulatory frameworks to enable achievement of these targets and

Table 1 India's RPO trajectory till 2029–2030

Year	Wind (%)	Hydro (%)	Distributed RE (%)	Other RE (%)	Total RE (%)
2024–2025	0.67	0.38	1.5	27.35	29.91
2025–2026	1.45	1.22	2.1	28.24	33.01
2026–2027	1.97	1.34	2.7	29.94	35.95
2027–2028	2.45	1.42	3.3	31.64	38.81
2028–2029	2.95	1.42	3.9	33.10	41.36
2029–2030	3.48	1.33	4.5	34.02	43.33

capacity addition through targeted programs (Fig. 3). The NAPCC established a foundation for these developments, which were followed by ambitious targets such as achieving 50% cumulative installed capacity from non-fossil-based resources by 2030 and Net Zero by 2070 [26]. These programs address various areas of RE development, including

1. Solar: The Jawaharlal Nehru National Solar Mission (JNNSM) was launched in 2010 as one of the missions under NAPCC with the objective to enhance solar power in the total energy mix [30]. Aligning with JNNSM to achieve grid parity by 2022, programs such as the scheme for the development of Solar Parks and Ultra Mega Solar Power Projects (UMSPPs) were launched [31, 32]. Furthermore, Roof-Top Solar (RTS) schemes, 2015 and in 2019 phase 2 of the same known as SRISTI, was launched, which aimed to promote targeted capacity of 40 GW from rooftop solar [33, 34]. Among the off-grid projects, Pradhan Mantri Kisan Urjas Suraksha evam Utthan Mahabhiyan (PMKUSUM) scheme was launched in March 2019, to promote off-grid solar projects by installing solar-powered agricultural pumps and standalone PV systems [33]. Under this scheme, the targeted solar capacity addition is 30.80 GW by 2026.

By December 2023, the achieved total solar capacity was 75.57 GW, with rooftop solar reaching only 11.08 GW, falling short of the targeted 100 GW and 40 GW set for 2022, respectively. This failure to achieve the set targets led to questions and concerns about the policy and regulatory designs and implementations on the ground level. Further in January 2024, to accelerate the adoption of rooftop solar and achieve the further upcoming RE targets, Roof-Top Solar (RTS) scheme has been renamed as PM Surya Ghar Muft Bijli Yojana which aims to accelerate the adoption of roof top solar and help improve the energy dependence [33, 34].

2. *Wind:* Wind energy witnessed a significant interest of investors due to favorable fiscal incentives through accelerated depreciation and attractive FiT. Although delayed adoption of reverse bidding for wind power led to competitive price discovery (~Rs3/kWh) as compared to FiT of Rs 5.6/kWh. However, this policy has since been reversed, possibly due to the pressure mounted by interest groups, and it has been recently reported to be reinstated. The significant gap between the competitively discovered tariff and the regulated one highlights the cost inefficiencies embedded in the FiT mechanism, for the wind energy sector [35].

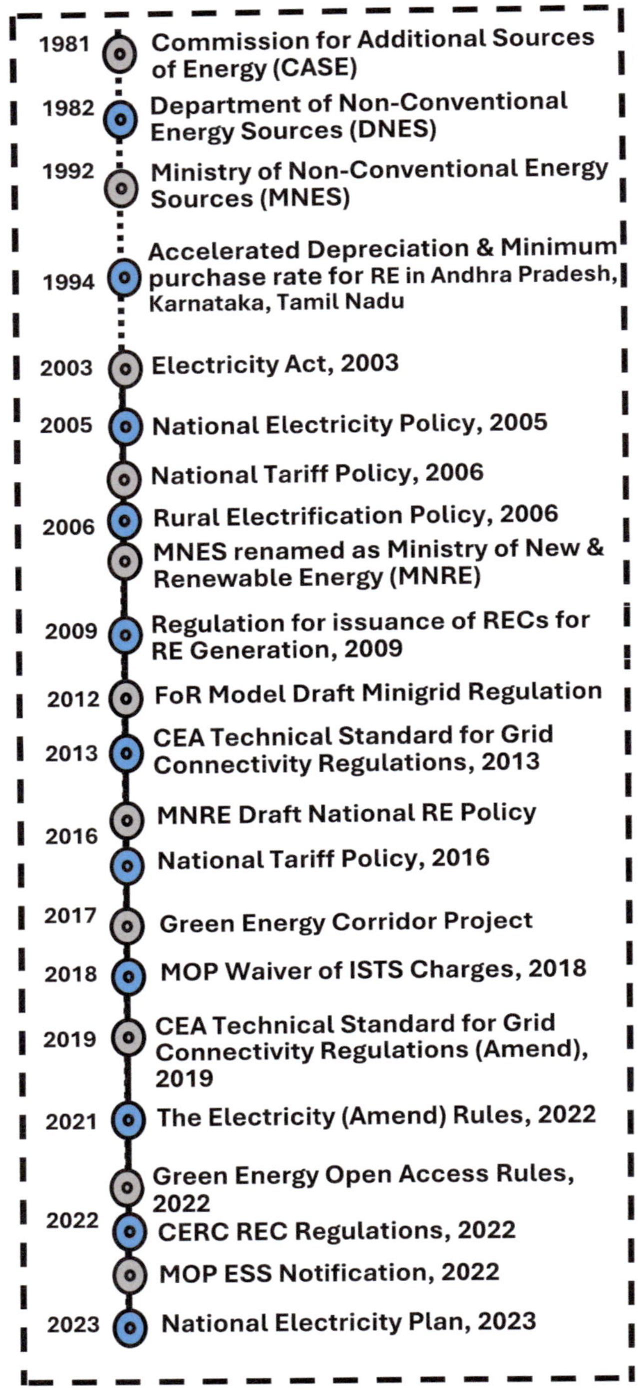

Fig. 3 Timelines of RE policy and regulations

Although offshore wind provides higher CUF, it comes at a much higher capital cost. The National Offshore Wind Energy Policy 2015 aims to provide a framework for optimum exploitation of offshore wind energy, promoting investment for installations in Exclusive Economic Zone (EEZs) [36]. Similarly, MNRE issued the National Wind-Solar Hybrid Policy in 2016, to maximize utilization of transmission systems and land use and minimize the variability in the system, hence achieving better grid stability [36].

3. *Bioenergy:* MNRE's National Bioenergy Programme 2021 promotes cleaner fuel sources and reduces fossil fuel dependence through sub-schemes for waste to energy, biomass, and biogas, including setting up large biogas plants using urban, industrial, and agricultural wastes [38]. The program emphasizes R&D on advanced biofuel and the manufacturing of pellets and briquettes [37], leading to financial support for families in rural areas to set up medium-sized biogas plants and promoting biomass-based cogeneration, until 2025–2026 [39]. Additionally, the CERCs Tariff Regulations, 2019, incentivize biomass blending with coal for thermal power generation, further encouraging biomass utilization and reducing stubble burning [40].

4. ***Transmission:*** The transmission is an essential infrastructural requirement for enhancing RE adoption, as most RES are geography specific and need connectivity up to the load centers. The Green Energy Corridors (GEC) scheme tackled the crucial challenge of evacuating RE power from generation-rich states to another state, i.e., Inter-State Transmission System (ISTS) and within the state boundaries, i.e., Intra-State Transmission System (InSTS), ensuring seamless integration into the grid [41, 42].

GEC InSTS Phase-I has established 8,940 ckm of transmission lines and 21,293 MVA substations [41] and GEC ISTS Phase-I has established 3200 ckm of interstate transmission lines and 17,000 MVA substations, while InSTS Phase-II has a target of 10,753 ckm of transmission lines and 27,546 MVA substations, to evacuate 20 GW RE power by 2025–2026 [41].

5. ***Energy Access:*** To improve energy access in rural and remote areas, India launched a series of initiatives. The Rajiv Gandhi Grameen Vidyutikaran Yojana (RGGVY) 2005 laid the groundwork for electrification, while Deen Dayal Upadhyaya Grameen Jyoti Yojana (DDUGJY) 2014 built upon this by focusing on strengthening the distribution network. Further advancements came through SERCs with their mini/microgrid policies and regulations, integrating advanced RE technologies like Battery Energy Storage Systems (BESS). The recent initiative, the Pradhan Mantri-Janjati Adivasi Nyaya Maha Abhiyan (PM-JANMAN) scheme (2024) targets the specific challenge of electrifying 1 lakh un-electrified tribal households, uplifting Particularly Vulnerable Tribal Groups (PVTGs) with a financial outlay of Rs. 24,000 Crore [43].

Additionally, policy initiatives were taken to develop advanced technology such as green hydrogen and EV while strengthening Research and Development (R&D). National Green Hydrogen Mission 2023 was launched with objective to make India a green hydrogen hub, with a target of 5 million tons of annual production by 2030 [44].

In addition to these programs and schemes, the next subsection delves into the financial drivers leading to the growing capacity addition of RES.

3 Financial Drivers for RE Growth

To enable adoption of solar energy under "unviable" conditions especially that for the public sector, a Viability Gap Funding (VGF) was introduced in 2014 as a part of JNNSM Phase-II, Batch-I [46]. VGF continued for subsequent batches, boosting RE growth, but due to increased competition, these generation-based incentives were withdrawn in 2017–2018. However, later MNRE approved the provision of VGF for setting up solar photovoltaic (SPV) projects for self-use or use by government entities under the Central Public Sector Undertaking (CPSU) scheme phase-II [47].

The RE sector attracted significant investment through Foreign Direct Investment (FDI) policy, where 100% FDI is permitted in the sector without prior government approval. This has attracted USD 6137.39 million FDI equity investment in the RE sector between 2020 and September 2023 [48]. Interstate Transmission System (ISTS) charges for solar, wind, Pumped Storage Hydro (PSH), and BESS projects set up before 2025 are waived off for the period of 25 years. Similarly, complete ISTS charges are waived off for offshore wind projects and green hydrogen plants/green ammonia plants commissioned before 2032 and 2030, respectively, for a period of 25 years [49].

Due to decline in capital cost and competitive bidding under Section 63 of the EA 2003, solar tariffs dropped from Rs. 6.17/kWh in FY-14 to Rs. 1.99/kWh in FY-21, while wind tariffs fell from Rs. 5.92/kWh in FY-16 to Rs. 3/kWh in FY-21. Notably, the wind energy regime transitioned from FiT to competitive bidding in December 2017, contributing to the sharp decline in wind tariffs from FY-18 (Rs. 5.4/kWh) to FY-19 (Rs. 2.83/kWh) (Fig. 4) [50].

The "Production Linked Incentive (PLI) Scheme," launched in April 2021, aims to support and encourage domestic manufacturing of high-efficiency solar modules. This scheme, implemented in two tranches with a total investment of Rs. 24,000 Crore, aims to reduce reliance on imports and establish India as a solar manufacturing powerhouse [47, 51].

4 Challenges to Achieve RE Targets

The current share of the installed capacity of RE is 31.43% (135.12 GW) as of January 2024 which is far away from 2022 target of 175 GW. The challenges persist for the fulfillment of its clean energy and RE development targets, primarily due to lack of institutional readiness, technical, and operational challenges with high VRE. Moreover, poor financial status of discoms and growing rooftop solar capacity poses greater stress on DISCOM financers due to the need of cross subsidy.

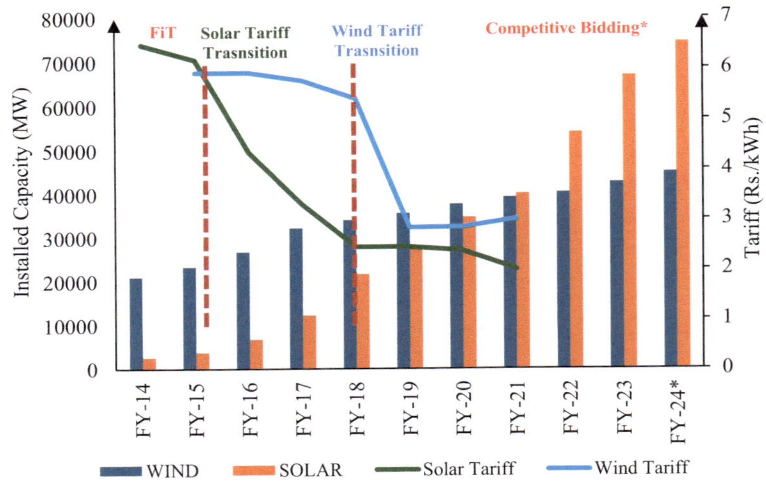

Fig. 4 RE tariff trends in India [35]

Despite the installed capacity of 31.43%, the actual generation from these RES is merely 13%, contrasting with the significant 73.92% share of generation from thermal plants in the year 2022–2023. At state level, RE has crossed over 50% (for example, in Karnataka) with few instances of more than 100% RE generation than the demand met within the state. The variable RES, i.e., solar and wind energy, exhibit variability and uncertainty in their generation across time blocks and seasons presenting a greater challenge for the system operator. These variable RE (VRE) sources cause intermittency, grid stability issues, and energy reliability concerns with increasing RE share [52]. The conventional plants, primarily thermal generators, need to ramp up/down to meet variability of generation over the day (Fig. 5), the intermittency demands greater flexibility from the power system constituents [53].

Additionally, technological advancements, the adoption of EVs, and the production of hydrogen through electrolysis would impact the demand pattern in the future [54]. Importance of visibility of behind the meter generation and trading has been highlighted in new compliance for energy technology like EV and green hydrogen [55]. With evolving power systems and an increasing share of RES, there is a need for enhanced power system flexibility to address the rising challenges of RE integration. This will enable a robust power system that can manage the variability and uncertainty of demand and supply reliably and cost-effectively across all relevant timescales. Enhanced system flexibility can be achieved through demand response and behind the meter generation plus storage on the demand side, and flexible thermal plant operations and expanding the transmission network, on the supply side [45]. Improved

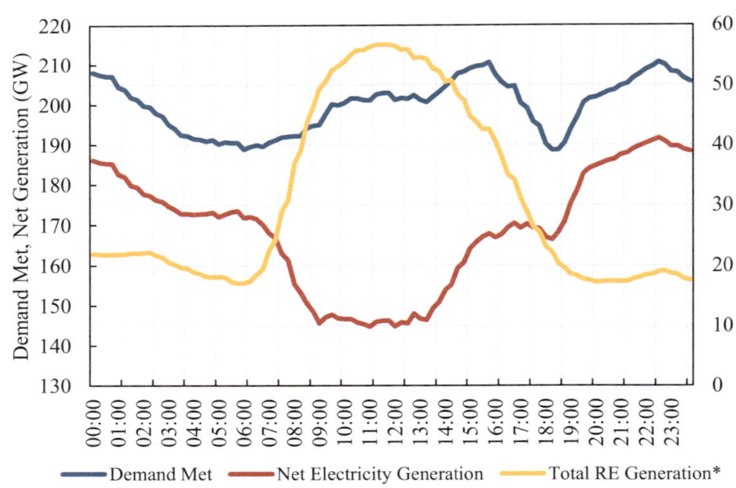

Fig. 5 Representation of nation-wide generation and demand changes in a day [53]

forecasting of demand and RE generation and transmission can also help address the flexibility needs of the power system [2, 56].

Policy and regulatory interventions are imperative for enhanced power system flexibility and facilitate greater RE adoption. These interventions include establishing clear standards for RE integration, system operation codes, and procedures for scheduling and dispatching power [22] (Fig. 5). Additionally, ESS, including BESS and PHS, play a critical role by mitigating the variability and enhancing RE adoption in the sector [56]. The EA 2003 allows distribution licensees to procure firm and dispatchable RE power. Round the Clock (RTC) RE plants with storage devices and wind–solar hybrid plants play important role in reduction of impacts of VRE. Additionally, market instruments like the Real-Time market (RTM) and Ancillary Service Market incentivize the development and deployment of dispatchable and fast-response technologies that can address system flexibility needs [57].

5 Conclusions and Way Forward

India's commitments to achieve its climate goals bank heavily on energy transition in the power sector, which has a greater role of RES. Apart from policy and regulatory push, institutional challenges across the sector need to be addressed to ensure that the targets can be achieved. The challenge of RE integration and high cost of storage to address variability and flexibility needs of the power sector would bring economies of electricity delivery to the forefront. Competitive bidding must remain the key root to encourage greater RE capacity in the country. This would also address rigidity of DISCOMs to embrace and support greater role of RES in the Indian power sector.

Even though most of the ambitious targets for RE addition were not achieved, new higher targets place a greater challenge for policymakers, regulators, and stakeholders. At state level, more or less similar scope of solar policy is "expected" to deliver greater result to achieve higher target, while the present version of the policy didn't deliver the outcome. This emphasizes the need to assess policy gaps and implementation challenges from the experience of existing policy [58].

Apart from technical and operational challenges, financial impacts on the distribution utilities and institutional rigidities would continue to demand innovative ideas riding policy and regulatory push. Agriculture solarizations and rooftop solar programs for subsidized households reduce the

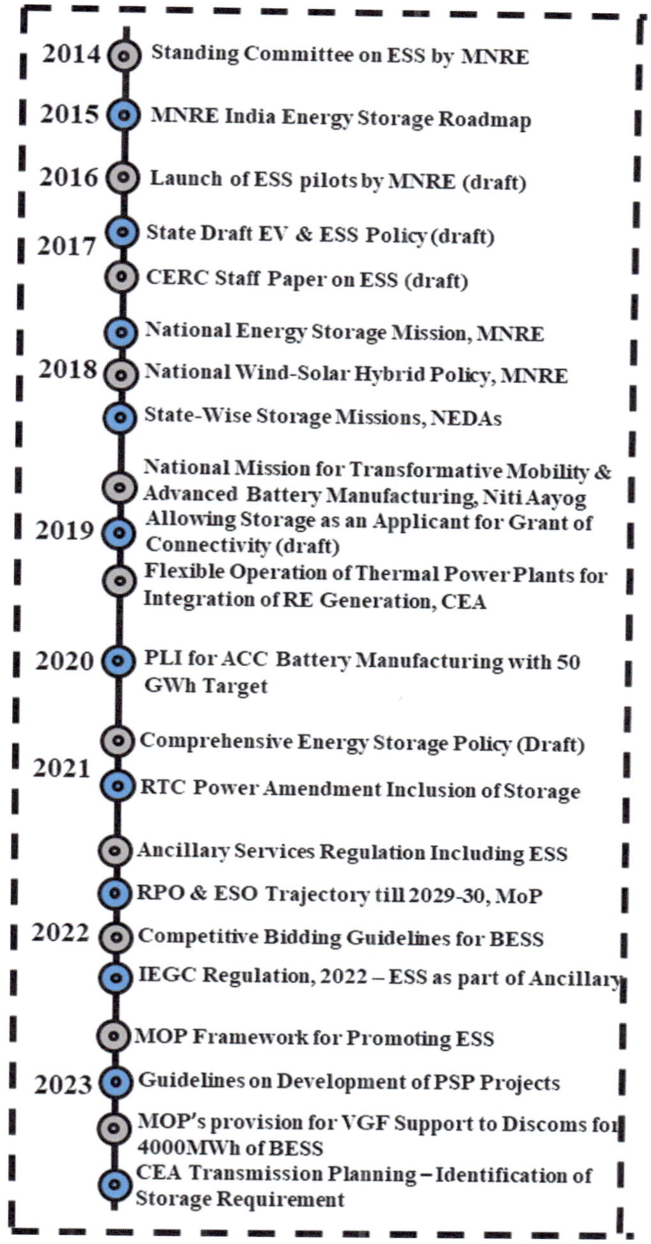

Fig. 6 Timelines of energy storage initiatives

power subsidy burdens on DISCOMs. Moreover, data visibility for behind the meter generation and better RE forecasting are required to address the issues due to high VRE in the systems.

Acknowledgements This research is supported by grants received from the "Power Sector Reform (PSR) Program—Phase II," funded by Government of the United Kingdom. The research has been conducted at the Center for Energy Regulation (CER), and Energy Analytics Lab (EAL) at Department of Management Science (DoMS), IIT, Kanpur. The author acknowledges the support and thanks the funding agencies for their generous support.

References

1. Yang J, Yu Y, Ma T, Zhang C, Wang Q (2021) Evolution of energy and metal demand driven by industrial revolutions and its trend analysis. Chin J Popul, Resour Environ 19(3):256–264. https://doi.org/10.1016/J.CJPRE.2021.12.028
2. Taibi E, et al (2018) Power system flexibility for the energy transition. https://doi.org/10.13140/RG.2.2.11150.61768
3. World Bank (2024) CO_2 emissions (metric tons per capita)|Data. World Bank Database. https://data.worldbank.org/indicator/en.atm.co2e.pc
4. Khare V, Khare CJ, Nema S, Baredar P (2022) Path towards sustainable energy development: Status of renewable energy in Indian subcontinent. Clean Energy Syst 3:100020. https://doi.org/10.1016/J.CLES.2022.100020
5. Energy and carbon emissions of buildings. Low Energy Low Carbon Arch, pp 41–56, May 2016. https://doi.org/10.1201/B19882-8
6. Shyam B, Kanakasabapathy P (2018) Renewable energy utilization in India-Policies, opportunities and challenges. In: Proceedings of 2017 IEEE international conference on technological advancements in power and energy, TAP Energy 2017, pp 1–6. https://doi.org/10.1109/TAPENERGY.2017.8397311
7. Agarwal N, Kumar A, Varun (2013) Optimization of grid independent hybrid PV-diesel-battery system for power generation in remote villages of Uttar Pradesh, India. Energy Sustain Dev. https://doi.org/10.1016/J.ESD.2013.02.002
8. Singh A (2006) Power sector reform in India: current issues and prospects. Energy Policy. https://doi.org/10.1016/J.ENPOL.2004.08.013
9. Singh A (2009) Informal markets for electricity: economics of lighting for hawkers in India. Int J Energy Manag. https://doi.org/10.1108/17506220910986824
10. Carrasco LM, Narvarte L, Martínez-Moreno F, Moretón R (2014) In-field assessment of batteries and PV modules in a large photovoltaic rural electrification programme. Energy 75:281–288. https://doi.org/10.1016/J.ENERGY.2014.07.074
11. Cust J, Singh A, Neuhoff K (2007) Rural electrification in India: economic and institutional aspects of renewables. SSRN. https://doi.org/10.2139/SSRN.2760810
12. Schmid G (2012) The development of renewable energy power in India: Which policies have been effective? Energy Policy. https://doi.org/10.1016/J.ENPOL.2012.02.039
13. Ministry of Power (MoP) (2024) The electricity act, 2003 ministry of law and justice (Legislative Department), 2003. https://cercind.gov.in/Act-with-amendment.pdf
14. Bhattacharyya SC (2005) The Electricity Act 2003: will it transform the Indian power sector? Utility Policy. https://doi.org/10.1016/J.JUP.2004.08.001
15. Singh A (2009) A market for renewable energy credits in the Indian power sector. Renew Sustain Energy Rev 13(3):643–652. https://doi.org/10.1016/J.RSER.2007.10.011
16. Singh A (2010) Economics, regulation, and implementation strategy for renewable energy certificates in India, India Infrastructure Report 2010, Oxford University Press
17. Gandhi HH, Hoex B, Hallam BJ (2022) Strategic investment risks threatening India's renewable energy ambition. Energ Strat Rev. https://doi.org/10.1016/J.ESR.2022.100921
18. Government of India (GoI) (2022) India's Updated First Nationally Determined Contribution Under Paris Agreement. Accessed: Mar. 10, 2024. https://unfccc.int/sites/default/files/NDC/2022-08/India%20Updated%20First%20Nationally%20Determined%20Contrib.pdf
19. Indian Parliamentary Database (2024) State-wise details of grid interactive renewable power installed capacity during the years 2009–2010. Accessed: Mar. 13, 2024. https://sansad.in/getFile/annex/220/Au89.htm?source=pqars
20. Central Electricity Authority (CEA) (2024) National power portal, ministry of power (MoP), government of India (GoI). Accessed: Feb. 27, 2024. https://npp.gov.in/dashBoard/gc-map-dashboard
21. Central Electricity Authority (CEA) (2024) Report on growth of electricity sector in India. Government of India (GoI). Accessed: Feb. 27, 2024. https://cea.nic.in/
22. Das P, Mathuria P, Bhakar R, Mathur J, Kanudia A, Singh A (2020) Flexibility requirement for large-scale renewable energy integration in Indian power system: Technology, policy and modeling options. Energ Strat Rev. https://doi.org/10.1016/j.esr.2020.100482
23. Energy efficiency & renewable energy management centre (EEREM), Government of NCT of Delhi, "Guideline for claim of Generation Based Incentive (GBI)." Accessed: Mar. 12, 2024. https://eerem.delhi.gov.in/sites/default/files/eerem/important-news/guidelines_on_disbursement_of_generation_based_incentive_gbi.pdf
24. CERC (2023) Central electricity regulatory commission (Indian Electricity Grid Code) regulations. Government of India. Accessed: Mar. 03, 2024. https://cercind.gov.in/Regulations/180-Regulations.pdf
25. Swamy D (2015) Competitive bidding for wind sector. Accessed: Mar. 03, 2024. http://www.rerc.rajasthan.gov.in/Orders/Order140.pdf
26. Government of India (GoI) (2009) National action plan on climate change. Accessed: Mar. 03, 2024. https://archivepmo.nic.in/drmanmohansingh/climate_change_english.pdf.
27. Government of India (GoI) (2024) India's intended nationally determined contribution. Accessed: Mar. 03, 2024. https://unfccc.int/sites/default/files/NDC/2022-06/INDIA%20INDC%20TO%20UNFCCC.pdf.
28. IRENA (2022) Renewable power generation costs in 2022. Accessed: Mar. 03, 2024. https://mc-cd8320d4-36a1-40ac-83cc-3389-cdn-endpoint.azureedge.net/-/media/Files/IRENA/Agency/Publication/2023/Aug/IRENA_Renewable_power_generation_costs_in_2022.pdf
29. Ackermann T, Söder L (2000) Wind energy technology and current status: a review. Renew Sustain Energy Rev. https://doi.org/10.1016/S1364-0321(00)00004-6
30. Jawaharlal Nehru national solar mission towards building solar India (2008). Accessed: Feb. 27, 2024. https://www.seci.co.in/upload/static/files/mission_document_JNNSM(1).pdf
31. MNRE (2022) Amendment in guidelines for disbursal of central grant for the development of Intra-State Transmission System under Green Energy Corridor in the states of Andhra Pradesh, Madhya Pradesh, Tamil Nadu, Himachal Pradesh. Accessed: Feb. 27, 2024. https://cdnbbsr.s3waas.gov.in/s3716e1b8c6cd17b771da77391355749f3/uploads/2022/12/2022122730.pdf
32. MNRE (2014) Implementation of a scheme for development of solar parks and ultra mega solar power projects. GoI. Accessed: Feb. 28, 2024. https://cdnbbsr.s3waas.gov.in/s3716e1b8c6cd17b771da77391355749f3/uploads/2022/12/2022122199-1.pdf
33. MNRE (2024) Guidelines for Implementation of Pradhan Mantri Kisan Urja Suraksha Evam Utthan Mahabhiyan (PM KUSUM). GoI. Accessed: Feb. 28, 2024. https://cdnbbsr.s3waas.gov.in/s3716e1b8c6cd17b771da77391355749f3/uploads/2022/12/2022122224-1.pdf
34. MNRE (2022) Guidelines on implementation of Phase-II of grid connected rooftop solar programme for achieving 40 GW capacity from Rooftop Solar by the year 2022. GoI. Accessed: Feb. 27, 2024. https://pmsuryaghar.gov.in/notification/Notification-18122017-152723.pdf
35. MNRE (2023) MNRE annual report 2020–2021. Accessed: Feb. 27, 2024. https://cdnbbsr.s3waas.gov.in/s3716e1b8c6cd17b771da77391355749f3/uploads/2023/08/202308311964406929.pdf

36. MNRE (2024) National wind-solar hybrid policy. Government of India (GoI). Accessed: Feb. 27, 2024. https://cdnbbsr.s3waas.gov.in/s3716e1b8c6cd17b771da77391355749f3/uploads/2022/12/2022122036.pdf
37. MNRE (2025) Guidelines for implementation of biomass programme for the period of FY 2021–2022 to 2025–2026, Government of India
38. MNRE (2022) Waste to energy programme. India. Accessed: Feb. 27, 2024. https://gobardhan.co.in/whats-new/Waste_to_Energy_Program%20revised%20guidelines.pdf
39. MNRE (2024) Biogas programme. Government of India (GoI). Accessed: Feb. 27, 2024. https://biogas.mnre.gov.in/public/downloadables/MNRE%20Biogas%20Prog%202021%20to%202026%20%20dated%2002_11_2022.pdf
40. CERC (2019) Terms and conditions of tariff regulations. GoI. Accessed: Mar. 13, 2024. https://cercind.gov.in/2019/regulation/Tariff%20Regulations-2019.pdf
41. MNRE (2024) InSTS GEC phase I achievements. Government of India (GoI). Accessed: Feb. 26, 2024. https://mnre.gov.in/gec-1/
42. Ministry of Power (MoP) (2024) Green energy corridor (GEC). GoI. Accessed: Feb. 26, 2024. https://powermin.gov.in/en/content/green-energy-corridor
43. MNRE (2024) Pradhan Mantri Janjati Adivasi Nyaya Maha Abhiyan (PM-JANMAN)," GoI. https://cdnbbsr.s3waas.gov.in/s3716e1b8c6cd17b771da77391355749f3/uploads/2024/02/202402021820513767.pdf
44. MNRE (2023) National green hydrogen mission (January). Accessed: Feb. 27, 2024. https://www.india.gov.in/spotlight/national-green-hydrogen-mission
45. Singh A (2021) MOP: national electricity policy, (2021) (draft). Regul Ints 4(1):2021
46. Aggarwal M, Arjun D (2018) State of the Indian renewable energy sector: drivers, risks, and opportunities. Counc Energy, Environ Water (September)1–48
47. MNRE (2019) CPSU Scheme Phase-II, Government of India (GoI), 2019. https://cdnbbsr.s3waas.gov.in/s3716e1b8c6cd17b771da77391355749f3/uploads/2023/08/2023080886.pdf
48. Ministry of Commerce and Industry (2020) Consolidated FDI policy. GoI. Accessed: Mar. 12, 2024. https://www.meity.gov.in/writereaddata/files/FDI-PolicyCircular-2020-29October2020_0.pdf
49. CEA (2024) India transmission portal: policy-ISTS waiver. Accessed: Feb. 26, 2024. https://indiatransmission.org/listdocuments/Policy/ISTS Waiver
50. CRISIL (2017) Competitive bidding set to transform wind power market. https://www.crisil.com/content/dam/crisil/our-analysis/views-and-commentaries/insights/CRISIL-Research-Insights-Competitive-bidding-a-blow-for-wind-power-sector-may2017.pdf
51. MNRE (2024) Production linked incentive (PLI) scheme (Tranche II) under national programme on high efficiency solar PV modules'. Government of India (GoI). Accessed: Feb. 27, 2024. https://cdnbbsr.s3waas.gov.in/s3716e1b8c6cd17b771da77391355749f3/uploads/2023/08/2023080863.pdf
52. Mohandes B, El Moursi MS, Hatziargyriou N, El Khatib S (2019) A review of power system flexibility with high penetration of renewables. IEEE Trans Power Syst. https://doi.org/10.1109/TPWRS.2019.2897727
53. Singh A (2022) All India generation and demand, power chronicle. Energy Analytics Lab (EAL) 5(2)
54. Zhang Q, Yan J, Gao HO, You F (2023) A systematic review on power systems planning and operations management with grid integration of transportation electrification at scale. Adv Appl Energy 11:100147. https://doi.org/10.1016/J.ADAPEN.2023.100147
55. POSOCO (2019) Analysis of its impact on Indian power system-a report power system operation corporation Ltd. Accessed: Feb. 26, 2024. https://cea.nic.in/old/reports/others/thermal/trm/flexible_operation.pdf
56. Das P, Mathuria P, Bhakar R, Mathur J, Kanudia A, Singh A (2020) Flexibility requirement for large-scale renewable energy integration in Indian power system: technology, policy and modeling options. Energy Strat Rev 29. https://doi.org/10.1016/j.esr.2020.100482.
57. Kumar TB, Singh A (2021) Ancillary services in the Indian power sector—a look at recent developments and prospects. Energy Policy. https://doi.org/10.1016/J.ENPOL.2020.112020
58. Anoop Singh (2022) UP solar policy, (Draft). Regulatory Insights 5(2)

MIX
Papier aus verantwortungsvollen Quellen
Paper from responsible sources
FSC® C105338

If you have any concerns about our products,
you can contact us on
ProductSafety@springernature.com

In case Publisher is established outside the EU,
the EU authorized representative is:
Springer Nature Customer Service Center GmbH
Europaplatz 3, 69115 Heidelberg, Germany

Printed by Libri Plureos GmbH
in Hamburg, Germany